Studies in Computational Intelligence

Volume 705

Series editor

Janusz Kacprzyk, Polish Academy of Sciences, Warsaw, Poland
e-mail: kacprzyk@ibspan.waw.pl

About this Series

The series "Studies in Computational Intelligence" (SCI) publishes new developments and advances in the various areas of computational intelligence—quickly and with a high quality. The intent is to cover the theory, applications, and design methods of computational intelligence, as embedded in the fields of engineering, computer science, physics and life sciences, as well as the methodologies behind them. The series contains monographs, lecture notes and edited volumes in computational intelligence spanning the areas of neural networks, connectionist systems, genetic algorithms, evolutionary computation, artificial intelligence, cellular automata, self-organizing systems, soft computing, fuzzy systems, and hybrid intelligent systems. Of particular value to both the contributors and the readership are the short publication timeframe and the worldwide distribution, which enable both wide and rapid dissemination of research output.

More information about this series at http://www.springer.com/series/7092

Arun Kumar Sangaiah · Ajith Abraham
Patrick Siarry · Michael Sheng
Editors

Intelligent Decision Support Systems for Sustainable Computing

Paradigms and Applications

 Springer

Editors
Arun Kumar Sangaiah
School of Computing Science
  and Engineering
VIT University
Vellore, Tamil Nadu
India

Ajith Abraham
Scientific Network for Innovation
  and Research Excellence
Machine Intelligence Research Labs
Auburn, WA
USA

Patrick Siarry
Faculté des Sciences et Technologie
Université Paris-Est Créteil Val-de-Marne
Créteil Cedex
France

Michael Sheng
Department of Computing
Macquarie University
Sydney, NSW
Australia

ISSN 1860-949X   ISSN 1860-9503 (electronic)
Studies in Computational Intelligence
ISBN 978-3-319-85078-8   ISBN 978-3-319-53153-3 (eBook)
DOI 10.1007/978-3-319-53153-3

Printed on acid-free paper

This Springer imprint is published by Springer Nature
The registered company is Springer International Publishing AG
The registered company address is: Gewerbestrasse 11, 6330 Cham, Switzerland

Foreword

The fast development of Internet technologies has raised the need for sustainable computing, namely energy-aware management of computing resources to efficiently control and reduce ecological and societal impacts. Most prominently, the rapid development and the world-wide adoption of cloud computing and data centers are critical challenges to sustainable computing in the twenty-first century. The aim is therefore to find trade-offs in the digital ecosystems, which support computation and data storage at large scale while ensuring eco-friendly computing.

The book *"Intelligent Decision Support Systems for Sustainable Computing"* addresses research and development issues in sustainable computing through a variety of evolutionary computational paradigms and applications. This book is a significant collection of 14 chapters covering various computational intelligence techniques, as well as their applications in sustainable computing that have emerged in the recent decade. This book provides an excellent platform to review various areas of intelligence decision support and analytics in depth, and caters for the needs of researchers and practitioners in the field. The multi-disciplinary nature of the sustainable computing spanning fields of computer science, optimization, soft computing, electrical engineering, etc. is sought in the book by a variety of paradigms, such as neuro-fuzzy, genetic algorithms, and optimization techniques, which are focused on real-world decision-making analysis, modeling, optimization, and control problems arising in the area of sustainable computing.

To my knowledge, this is the first attempt to comprehensively use evolutionary computation and provide an intensive and in-depth coverage of the key subjects in the fields of evolutionary computational intelligence approaches to sustainable computing. This book is an invaluable, topical, and timely source of knowledge in the field, which would serve nicely as a major textbook for several courses at both undergraduate and postgraduate levels. It can also serve as a key reference for scientists, professionals, researchers, and academicians, who are interested in new challenges, theories, practice, and advanced applications of the specific area of sustainable computing.

I am happy to inform the readers about the interesting findings of *intelligent decision support systems* via computational intelligence approaches on sustainable computing. This book is a main step in this field's maturation and will serve to unify, advance, and challenge the scientific community in many important ways. This book provides a valuable contribution on sustainable computing by covering the necessary components such as energy efficiency and natural resource preservation and emphasizes the role of ICTs in achieving digital eco-system design and operation objectives.

I am happy to commend the editors and authors on their accomplishment, and wish that the readers find the book useful and a source of inspiration in their research and professional activity.

December 2016   Fatos Xhafa
                                                      Professor Titular d'Universitat
                                          Departament de Ciències de la Computació
                                              Universitat Politècnica de Catalunya
                                                                        Barcelona
                                                                            Spain

Preface

Sustainable computing has been extended to become a significant research area that covers the fields of computer science and engineering, electrical engineering, and other engineering disciplines. Recently, we have been witnessing from adequate literature of sustainable computing that includes energy efficiency and natural resource preservation and emphasize the role of ICT (information and communication technology) in achieving system design and operation objectives. The energy impact/design of more efficient IT infrastructure is a key challenge for organizations to realize new computing paradigms.

On the other hand, the uses of computational intelligence (CI) techniques for intelligent decision support can be exploited to originate effectual computing systems. CI consists of various branches that are not limited to expert systems (ES), artificial neural networks (ANN), genetic algorithms (GA) and fuzzy logic (FL), knowledge-based systems, and various hybrid systems, which are combinations of two or more of the branches. The intention of this book is to explore sustainability problems in computing and information processing environments and technologies at the different levels of CI paradigms. Moreover, this edited volume is to address the comprehensive nature of sustainability and to emphasize the character of CI in modeling, identification, optimization, prediction, forecasting, and control of complex systems.

The chapters included in this book focus on addressing latest research, innovative ideas, challenges, and CI solutions in sustainable computing. Moreover, these chapters specify novel in-depth fundamental research contributions from a methodological/application perspective in accomplishing sustainable lifestyle for society. This book provides a comprehensive overview of constituent paradigms underlying evolutionary computational intelligence methods, which are illustrating more attention to sustainability computing problems as they evolve. Hence, the main objective of the book is to facilitate a forum to a large variety of researchers, where decision-making approaches under CI paradigms are adapted to demonstrate how the proposed procedures as well as sustainability computing problems can be handled in practice.

Need for a Book on the Proposed Topics

Over the recent decades, many of useful methods have been proposed to solve organizational decision-making problems. Computational Intelligence paradigm in intelligence decision support systems has fostered a broad research area, and their significance has also been clearly justified at many applications. This volume addresses a wide spectrum of CI paradigms, making decisions of an industry or organization happened at all the levels of sustainable challenges.

This volume aims to provide relevant theoretical frameworks and the latest empirical research findings in the area. The CI approaches applied to sustainable computing decision-making systems usually have received more attention in recently published volumes. Based on this context, there is need envisioning for a key perspective into current state of practice of computational intelligence techniques. Consequently, to address the predicative analysis of sustainable computing problems including energy efficiency and natural resource preservation, load distribution strategy has been addressed in this book.

Solutions for these problems have been effectively handled through wide range of algorithmic and computational intelligence frameworks, such as optimization, machine learning, decision support systems, and meta-heuristics. The main contributions to this volume address sustainability problems in computing and information processing environments and technologies, and at various levels of the computational intelligence process.

Organization of the Book

This volume is organized into 14 chapters. A brief description of each chapter is given as follows:

Chapter "Intelligent Decision Support Systems for Sustainable Computing" gives an overview of computational intelligence paradigms in intelligent decision support and analytics for sustainable computing.

The editors briefly describe the various sustainability problems in computing and information processing environments and technologies, and at various levels of the computational intelligence process. The overall aim of the chapter is to address the convergence of CI methodologies in sustainable computing.

Chapter "A Genetic Algorithm Based Efficient Static Load Distribution Strategy for Handling Large-Scale Workloads on Sustainable Computing Systems" covers an efficient processor availability-aware scheduling model to optimize the energy efficiency of heterogeneous sustainable computing systems. Using this model, the authors design a genetic algorithm-based global optimization strategy to derive an optimal load partition together with an optimal distribution sequence.

Chapter "Efficiency in Energy Decision Support Systems Using Soft Computing Techniques" presents the problem of energy demand forecasting, and decision making has been investigated. The authors propose an integrated decision support system that involves Adaptive Neuro-Fuzzy Systems (ANFIS), Neural Networks (NN), and Fuzzy Cognitive Maps (FCM), along with Econometric Models (EM) in a hybrid fashion to predict energy consumption and prices.

Chapter "Computational Intelligence Based Heuristic Approach for Maximizing Energy Efficiency in Internet of Things" presents computational intelligence-based heuristic approach for maximizing energy efficiency in the Internet of Things (IoT). The authors present the Modified Multi-objective Particle Swarm Optimization (MMOPSO) algorithm based on the concept of dominance to solve the mobile cloud task scheduling problem. Overall, this chapter explores IoT and cloud computing as well as their symbiosis based on the common environment of distributed processing.

Chapter "Distributed Algorithm with Inherent Intelligence for Multi-cloud Resource Provisioning" illustrates distributed algorithms with inherent intelligence for multi-cloud resource provisioning. The authors introduce the substantial ranking method for elastic and inelastic tasks scheduler to support the heterogeneous requests and resources in multi-cloud environments.

Chapter "Parameter Optimization methods Based on Computational Intelligence Techniques in Context of Sustainable Computing" provides a comprehensive study of parameter optimization. Accuracy of any computational intelligence (CI) algorithm is highly dependent on optimal settings of parameters. The authors have discussed how parameter setting affects the performance and robustness of evolutionary algorithms.

Chapter "The Maximum Power Point Tracking Using Fuzzy Logic Algorithm for DC Motor Based Conveyor System" presents to design a conveyor belt system driven by DC motor whose speed is controlled by solar-powered converter operating at maximum power point (MPPT). In this chapter, the authors have compared the proposed MPPT algorithm and fuzzy approach with existing algorithms such as perturb, observe and incremental conductance.

Chapter "Differential Evolution Based Significant Data Region Identification on Large Storage Drives" emphasizes on the identification of data relevant sector regions in digital hard disk drives (HDD) using computationally intelligent differential evolution (DE) algorithm to accelerate the overall digital forensic (DF) process. In this chapter, the authors have presented a new trade-off rectangle to reveal present requirements toward the development of existing DF facilities. The chapter also proposes a methodology that extract data storage pattern using storage drive's structural information and DE algorithm that can help investigator in planning further course of action.

Chapter "A Fuzzy Based Power Switching Selection for Residential Application to Beat Peak Time Power Demand" introduces a fuzzy-based power switching selection for residential application to beat peak time power demand. The authors have designed the hybrid system using solar and main power as sources. The fuzzy logic algorithm has been used to select the power source based on the peak and off peak time, and utilized power level and availability of power source to meet the power demand during the peak time and off peak time.

Chapter "Energy Saving Using Memorization: A Novel Energy Efficient and Fault Tolerant Cluster Tree Algorithm for WSN" proposes Energy Saving Using Memorization (ESUM)—a novel energy-efficient and fault tolerant algorithm for cluster tree-based routing—that uses energy conservation to enhance network longevity, by using saved results and avoiding re-elections after each round. The authors presents a comparison of various cluster tree protocols and proposes a novel algorithm ESUM that leads to increase in network lifetime by reducing the communication overhead in cluster heads (CH) re-election.

The main objective of this Chapter "Analyzing Slavic Textual Sentiment Using Deep Convolutional Neural Networks" is to give a clear understanding of the position of low-resource languages and propose a direction for sustainable development of language technologies illustrated using convolutional neural networks for textual sentiment analysis. The authors have proposed a system which is based on supervised learning and can be quickly adapted to use simple text, circumventing the need for more intricate features.

Chapter "Intelligent Decision Support System for an Integrated Pest Management in Apple Orchard" presents hybrid case-based reasoning computational intelligence for pest management in apple production. In agriculture, intelligent decision support systems (IDSSs) have been used for the optimization of a number of planning and decision-making challenges under variable constraints based on noisy data. This chapter describes an IDSS to implement and optimize pest and disease protection decision-making processes within temperate regions of India; develops hybrid algorithm using case-based reasoning (CBR) and database technology, and implements the same using Web-based client server architecture.

Chapter "Analysis of Error Propagation in Safety Critical Software Systems: An Approach Based on UGF" presents a computational intelligence (CI)-based approach to compute the error inclusion in the output of the selected safety critical system. The authors address a novel Error Propagation Metric called Safety Metric SM_{EP}, which can be characterized depending on the performance rate of the software module. Through this, the performance distribution of system modules and the system with respect to safety metric SM_{EP} has been quantified.

Chapter "A Framework for Analyzing Uncertainty in Data Using Computational Intelligence Techniques" presents the design of an efficient classifier that handles ambiguity and vagueness in medical datasets for better diagnosis of illness. Moreover, the chapter authors investigate a rule-based fuzzy-rough classifier for analyzing uncertainty in medical dataset.

Audience

The intended audience of this book includes scientists, professionals, researchers, and academicians, who deal with the new challenges and advances in the specific areas mentioned above. Designers and developers of applications in these fields can learn from other experts and colleagues through studying this book. Many universities have started to offer courses on computational intelligence (CI), sustainable computing on the graduate/postgraduate level in information technology and management disciplines. This book starts with an introduction to sustainable computing and CI paradigms, hence suitable for university level courses as well as research scholars. Their insightful discussions and knowledge, based on references and research work, will lead to an excellent book and a great knowledge source.

Vellore, India

Arun Kumar Sangaiah

Auburn, WA, USA

Ajith Abraham

Créteil Cedex, France

Patrick Siarry

Sydney, Australia

Michael Sheng

Acknowledgements

The editors would like to acknowledge the help of all the people involved in this project and, more specifically, to the authors and reviewers that took part in the review process. Without their support, this book would not have become a reality.

First, the editors would like to thank each one of the authors for their contributions. Our sincere gratitude goes to the chapter's authors who contributed their time and expertise to this book.

Second, the editors wish to acknowledge the valuable contributions of the reviewers regarding the improvement of quality, coherence, and content presentation of chapters. We deeply appreciate the comments of the reviewers who helped us to refine the context of this book. Most of the authors also served as referees; we highly appreciate their double task.

Finally, our gratitude goes to all of our friends and colleagues, who were so generous with their encouragement, advice, and support.

Vellore, India   Arun Kumar Sangaiah
Auburn, WA, USA   Ajith Abraham
Créteil Cedex, France   Patrick Siarry
Sydney, Australia   Michael Sheng

Contents

Intelligent Decision Support Systems for Sustainable Computing

Arun Kumar Sangaiah, Ajith Abraham, Patrick Siarry
and Michael Sheng

Abstract In sustainable computing, Intelligent Decision Support Systems (IDSS) has been adopted for prediction, optimization and decision making challenges under variable number constraints based on un-structured data. The traditional systems are lack of efficiency, limited computational ability, inadequate and impreciseness nature of handling sustainable problems. Despite, Computational Intelligence (CI) paradigms have used for high computational power of intelligence system to integrate, analyze and share large volume of un-structured data in a real time, using diverse analytical techniques to discover sustainable information suitable for better decision making. In addition, CI has the ability to handle complex data using sophisticated mathematical models, analytical techniques. This chapter provides a brief overview of computational intelligence (CI) paradigms and its noteworthy character in intelligent decision support and analytics of sustainable computing problems. The objective of this chapter is to study and analyze the effect of CI for overall advancement of emerging sustainable computing technologies.

Keywords Sustainable computing · Computational intelligence · Intelligent decision support systems

A.K. Sangaiah (✉)
School of Computing Science and Engineering, VIT University, Vellore,
Tamil Nadu, India
e-mail: arunkumarsangaiah@gmail.com

A. Abraham
Machine Intelligence Research Labs, Auburn, USA

P. Siarry
Université Paris-Est Créteil Val-de-Marne, Créteil, France

M. Sheng
Department of Computing, Macquarie University, Sydney, NSW 2109, Australia

© Springer International Publishing AG 2017
A.K. Sangaiah et al. (eds.), *Intelligent Decision Support Systems*
for Sustainable Computing, Studies in Computational Intelligence 705,
DOI 10.1007/978-3-319-53153-3_1

1 Introduction

Sustainable Computing is a standard that encompasses on a wide variety of policies, procedures, programs, and attitudes that execute the length and breadth and make use of information technologies. Sustainable Computing has been played a key role in industries, society, product and services such as sustainable energy, tools, process, and supply chains. Subsequently, we might be consider a sustainable computing from the perspective of green computing (maximize the energy efficiency, recyclability). Sustainable ICTs (Information and Communication Technology) and management of green computing will leads to significant benefit in energy efficiency. Due to globalization has caused most of the developed countries making an initiative to step rise in energy consumption. Moreover, sustainable IT green computing green strategies will have a significant impact and environment benefits on innovate, create value and build a competitive advantage. Further, the green IT will gain numerous advantages: (a) reduction of power and energy consumption, (b) improved resource utilization, (c) better operational efficiency, (d) reduce cost, fulfil the compliance and regulatory requirements and so on. Consequently, lot of research has been produced towards the development of energy models via Internet of Things (IoT), smart sensors, context-aware systems, pervasive, cloud computing systems, that try to optimize environmental sustainability. Furthermore, only limited research studies have investigated Computational Intelligence (CI) paradigms and their applications in building and predicting sustainable energy system. So, based on this context, this book have addressed the various CI prediction methodologies for the sustainable computing and analytics has been addressed in this chapter and this is also the focus of this edited book.

2 Intelligent Decision Support Computational Intelligence Paradigms

Computational Intelligence and its decision support methodologies have the power to attain knowledge about a specific task from given data. A system which is called computationally intelligent if it handles with structured/un-structured data during the decision making process of any application. This book mainly has designed the CI paradigms to solve complex real world sustainable computing problems. Moreover, the intelligent decision support CI paradigms which are extract the historical data in order to predict the future sustainable problems.

CI methodologies are of various divisions that are not limited to, Neuro computing, evolutionary computing, granular computing and artificial Immune system and so on. Moreover, CI mainly represents the integration of the following five methods from the soft computing perspective: Fuzzy Logic, Artificial Neural

Networks, Evolutionary Computation, Learning Theory, and Probabilistic Methods. All these methods in fusion with one another helps the computer to solve a problem in the following way—(a) Fuzzy logic—understand natural language, (b) Artificial neural networks—to learn from experiential data by operating similar to the biological one, (c) Evolutionary computing—process of selection, (d) Learning theory—reasoning, and (e) Probabilistic methods—dealing with uncertainty imprecision.

Besides these main principles, there are some other approaches which include genetic algorithms, biologically inspired algorithms such as swarm intelligence and artificial immune systems. Recently there is an interest to fuse computational intelligence approaches with data mining, natural language processing, and artificial intelligence techniques. The detailed overview CI approaches on sustainable applications has been depicted in Fig. 1.

In sustainable computing techniques, we always require to solve several of optimization problems like design, planning and control, which are really hard. Traditional mathematical optimization techniques are limited capacity and computationally difficult. Recent advances in computational intelligence paradigms have resulted in an in-creasing number of optimization techniques such as (a) Nature-inspired computational approaches: ant colony optimization, bee algorithm, firefly optimization, bacterial foraging optimization, artificial immune system and etc. (b) evolutionary computational approaches: Genetic algorithms, particle swarm optimization etc. (c) logical search algorithms: tabu search, harmony search, cross entropy method and so on for effectively solve these complex problems. Basically, the algorithms which are depends on the principle of natural biological evolution and/or collective behaviour of swarm have addressed a promising performance that has been reported many literature studies. Similarly, the earlier studies [1–3] have used computational intelligence techniques for parameter optimization (parameter tuning and parameter control) for computational sustainability issues. Subsequently, the earlier researchers [4, 5] have compared the recent CI paradigms for parameter optimization on various performance indicators.

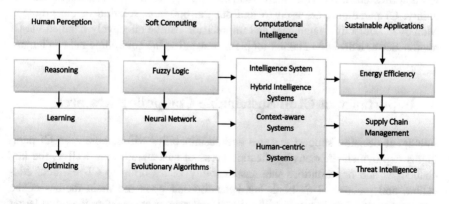

Fig. 1 Integration of CI paradigms on sustainable computing applications

3   CI Paradigms for Sustainable Applications

Computational Intelligence (CI) is the division of science and engineering where complex computational problems that are handled by modelling problems according to the natural and evolutionary intelligence, resulting in "intelligent systems". These intelligent systems comprises numbers of popular intelligent algorithms; artificial neural networks, artificial immune systems, evolutionary computation, fuzzy system and swarm intelligence. Consequently, these intelligent algorithms are being a part of Artificial Intelligence (AI). Alternatively, it is stated that Computational Intelligence (CI) is the successor of Artificial Intelligence (AI), where CI is analysis and study of adaptive mechanism to facilitate intelligent responses and is considered the sustainable computing and its applications. Recently, computational intelligence has emerged as a powerful methodology for revealing sustainable real-world challenging problems. The earlier study [6] have investigated the significance of CI, specifically neural networks in handling complexity and stochastic challenges with respect to smart grid system for addressing the new requirements of a sustainable global energy system.

Recently, there is an interest by researchers and practitioners have adopted CI paradigms for sustainable supply chain management. The previous studies [7, 8] have addressed the advances and applications of computational intelligence for sustainable supply chain planning in the context environmental sustainability (green design, green procurement, green production, green logistics, green packaging, green recycling). Subsequently, there are limited studies have focused on the supply chain management issue for supplier evaluation. Moreover, the studies have mostly focused on fuzzy multi-criteria decision making (MCDM) approaches such as Analytical Hierarchy Process, Analytical Network Process, DEMATEL, rough set theory for supplier evaluation. Further, these models have limited capacity (less robust) because computation in order quantity a specific supplier is not possible. To overcome such limitation hybrid fuzzy hybrid multi-criteria approaches has been investigated in the recent studies [9–11]. Still there are lots of research gaps needs to be addressed in the view meta-heuristic approaches for environmental sustainability. CI paradigms are used to find out the optimum solution of the problem. Moreover, subjective vagueness and imprecision can be effectively handled via CI approaches as a decision making tools for preventing the environmental challenges.

4   Importance of CI in Sustainable Computing Research

Intelligent decision support systems and sustainability will elaborate on CI paradigms deployment in many application areas of sustainability as well as the key challenges and opportunities that sustainability issues bring to CI research, education, and practice. This book has been focused upon the main themes at the intersection of CI and sustainability, but it will primarily concentrate on the larger contexts of sustainability, and on computing and sustainability, thereby setting the stage for extendable research needs to be carried out. The earlier study [12] has defined the sustainability as "Sustainable development is development that meets the needs of the present without compromising the ability of future generations to meet their own needs." Since there will be wide space for researchers large number sustainability challenges relating to energy, climate change, agriculture, transportation, disease, garbage and so on needs to be predict in advance in sustainable management. Similarity, CI paradigms and optimization algorithms have paved the way for effective design, and implementation of sustainable engineering applications. Typically, sustainable problems and challenges hard to handle with traditional approaches, to overcome such limitations, CI paradigms have take over a prominent role in planning, optimizing and forecasting sustainable systems. Typically, these methodologies would use of domain knowledge in order to obtain the required objectives. Hence, in the case that explicit domain knowledge is not presented, CI approaches could handle effectively with large raw numerical sensory data directly, process them, generate reliable and just-in-time responses, and have high fault tolerance.

The CI paradigms applied to sustainable computing and intelligent decision support systems have been paid more attention recently. Further, fusion of CI approaches such as neuro fuzzy approaches, artificial neural networks, evolutionary computation, swarm intelligence, rough sets can be incorporated to handle uncertainty and subjectivity in decision making process. However, the hybridization of CI techniques and optimization techniques has not been adequately investigated from the perspective of sustainable computing and analytics. Hence, there is an opportunity to address the emerging trends and advancement in sustainable application and process to by harnessing the power of computational intelligence. Future CI based applications would focus more on real-world problems that need a paradigm shift of paying attention to improving computational efficiency, understudying theoretical foundations and frameworks, and most essentially, supporting real decision-making in complex, uncertain application contexts. Taking ideas from CI which relate to sustainable computing and analytics to bringing into computational models would be helpful to researchers in this field to develop novel approaches and establish new research avenues to pursue.

5 Conclusion

In this chapter, the problem and challenges of sustainable computing in forecasting and decision making has been illustrated. The generic logic framework of the IDSS is highlighted, to outline the key functionalities of the CI on forecasting and the model repository as an open-ended subsystem including all relevant components for prediction. The research on development and application of CI paradigms and other meta-heuristic approaches can provide effective solutions for optimization problems, specifically, dealing with incomplete or inconsistent information and limited computational capability in handling sustainable problems. CI paradigms with relate to sustainable computing intelligent decision support and analytics bringing them into computational models can help researchers in this field to develop novel approaches and establish new research avenues to pursue.

References

1. A.E. Eiben, Z. Michalewicz, M. Schoenauer, J.E. Smith, Parameter control in evolutionary algorithms. IEEE Trans. Evol. Comput. **3**, 124–141 (1999)
2. Y.J. Zheng, S.Y. Chen, Y. Lin, W.L. Wang, Bio-inspired optimization of sustainable energy systems: a review. Math. Probl. Eng. (2013)
3. X.S. Yang, *Engineering Optimization: An Introduction with Metaheuristic Applications* (Wiley, 2010)
4. M.S. Norlina, P. Mazidah, N.M. Sin, M. Rusop, Application of metaheuristic algorithms in nano-process parameter optimization, in *2015 IEEE Congress on Evolutionary Computation (CEC)* (IEEE, 2015), pp. 2625–2630
5. N.M. Sabri, M. Puteh, M.R. Mahmood, An overview of gravitational search algorithm utilization in optimization problems, in *2013 IEEE 3rd International Conference on System Engineering and Technology (ICSET)* (IEEE, 2013), pp. 61–66
6. P.J. Werbos, Computational intelligence for the smart grid-history, challenges, and opportunities. IEEE Comput. Intell. Mag. **6**(3), 14–21 (2011)
7. R.J. Lin, Using fuzzy DEMATEL to evaluate the green supply chain management practices. J. Clean. Prod. **40**, 32–39 (2013)
8. K. Shaw, R. Shankar, S.S. Yadav, L.S. Thakur, Supplier selection using fuzzy AHP and fuzzy multi-objective linear programming for developing low carbon supply chain. Expert Syst. Appl. **39**(9), 8182–8192 (2012)
9. V. Jain, A.K. Sangaiah, S. Sakhuja et al., Supplier selection using fuzzy AHP and TOPSIS: a case study in the Indian automotive industry. Neural Comput. Appl. (2016). doi:10.1007/s00521-016-2533-z
10. O.W. Samuel, G.M. Asogbon, A.K. Sangaiah, P. Fang, G. Li, An integrated decision support system based on ANN and Fuzzy_AHP for heart failure risk prediction. Expert Syst. Appl. **68**, 163–172 (2017)
11. A.K. Sangaiah, J. Gopal, A. Basu, P.R. Subramaniam, An integrated fuzzy DEMATEL, TOPSIS, and ELECTRE approach for evaluating knowledge transfer effectiveness with reference to GSD project outcome. Neural Comput. Appl. (2015). doi:10.1007/s00521-015-2040-7
12. G.H. Brundtland (ed.), *Report of the World Commission on Environment and Development: Our Common Future*, United Nations (1987), http://www.un-documents.net/wced-ocf.htm

A Genetic Algorithm Based Efficient Static Load Distribution Strategy for Handling Large-Scale Workloads on Sustainable Computing Systems

Xiaoli Wang and Bharadwaj Veeravalli

Abstract A key challenge faced by large-scale computing platforms to go green is the effective utilization of energy at the various processing nodes. Most existing scheduling models assume that processors are able to stay online forever. In reality, processors, however, may have arbitrary unavailable time periods. Hence, if we inadvertently assign tasks to processors without considering the availability constraints, some processors would not be able to finish their assigned workloads. Thus all the unfinished workloads need to be reassigned to other available processors resulting in an inefficient time and energy schedule. In this chapter, we propose a novel *processor availability-aware divisible-load scheduling* model. Using this model, we design a time-efficient genetic algorithm based global optimization technique to derive an optimal load distribution strategy. Our experimental results show that the proposed algorithm adapts to minimize the processing time, hence the energy consumption too, by over 60% compared to other strategies.

Keywords Divisible load · Release time · Off-line time · Load distribution · Processor availability

1 Introduction

Modern large-scale computing platforms, such as networked computing systems and cloud computing, have imminent need to go green since they are severely constrained by energy related issues [1]. This is predominantly due to their heavy utilization of power and cooling resources which results in rapid energy consumption which in turn

X. Wang
School of Computer Science and Technology, Xidian University, Xi'an 710071, China
e-mail: wangxiaoli@mail.xidian.edu.cn

B. Veeravalli (✉)
Department of Electrical and Computer Engineering, National University of Singapore, Singapore 117576, Singapore
e-mail: elebv@nus.edu.sg

© Springer International Publishing AG 2017
A.K. Sangaiah et al. (eds.), *Intelligent Decision Support Systems for Sustainable Computing*, Studies in Computational Intelligence 705, DOI 10.1007/978-3-319-53153-3_2

7

imparts large carbon footprints on the environment [2]. Emerging sustainable computing technologies, which primarily aim at reducing massive energy consumption by developing certain ab initio computational and mathematical models, methods, and tools for resource allocation and task scheduling, are therefore gain significant interest to researchers and practitioners.

One of the key techniques to save energy is Dynamic Voltage Scaling (DVS). It exploits the hardware characteristics to save energy by degrading CPU voltage and operating frequency while keeping the processor to operate at a slow speed [3]. During the past decades, substantial energy-efficient scheduling strategies have been proposed for DVS-enabled systems [4–7]. These DVS-based techniques, however, may not applicable to virtualized environments where physical processors are shared by multiple virtual machines (VMs) as lowering the supply voltage will inadvertently affect the performance of VMs belonging to different applications [8]. Another promising approach to conserve energy is turning off idle computing nodes in a data center by packing the running VMs to as few physical servers as possible, often called VM consolidation [9]. However, live VM migration must be guaranteed and resources must be properly allocated in order to avoid severe performance degradation due to resource competition by co-located VMs. There are a large amount of studies on the migration strategies, concerning the issues of where, when, and how a VM should be migrated [10–13]. At the current stage, several management issues about VM consolidation still deserve additional investigations. For example, transferring large-sized data over the shared network link is a huge challenge, especially when several goals in terms of Service-Level Agreement (SLA) violation avoidance, minimum communication delay, high system throughput, and high quality of services have to meet [14].

While significant advancements have been made to minimize the energy consumption for sustainable computing, even stronger effort is needed to promote the effective utilization of energy at the various processing nodes. By "effective utilization" we mean that efforts need to be devoted to making compute platforms not just minimizing the energy consumption but also to make every amount of energy consumed for workload computation worthwhile. This is based on the fact that the actual energy consumed for workload computation might not be equal to the energy that is required for workload computation. For example, most existing scheduling models assume that the compute units, which are processors, are able to stay online and available forever. That is to say, it assumes that all processors remain idle at the beginning of the workload assignment and that they will be kept busy until the assigned workload fractions are completed. In reality, processors, however, may have arbitrary unavailable time periods. They may still be busy computing any previous workload even when a new workload arrives and may even get off-line before finish computing the currently assigned load. The time period between release time and off-line time of a processor is referred to as its available time period. Hence, if we inadvertently assign tasks to processors according to their computational capabilities without taking into account of the availability constraints, some processors would not be able to finish their assigned workloads. Thus all of the unfinished workloads need to be reassigned to other available processors resulting in an inefficient time and energy schedule. Therefore, designing an efficient load distribution strategy seems appropriate when one considers resource (processor) available times.

It is believed that workloads to be scheduled on Heterogenous Sustainable Computing Systems (HSCS) are quite large in size and possess computationally intensive CPU requirements; otherwise, one or a few processors should be enough for workload computation. Also, workloads should be partitionable so that they can simply be further divided into a number of load fractions and distributed to processors for independently parallel computing. Ideally, if a workload can be divided into an arbitrary number of load fractions such that there are no precedence relationships among these fractions, then we refer to it as a divisible load [15]. Actually, divisible loads exist in widely multiple real-world applications, such as real-time video encoding [16, 17], satellite image classification [18], signature searching in a networked collection of files [19], and so on. It may be noted that divisible load modelling can also be adopted for modern day Big Data processing when the requirements of processing demand homogeneous processing on the data.

There are considerable studies available on finding an optimal load distribution strategy for scheduling large-scale divisible loads on various distributed networks with different topologies, including linear networks [20], bus networks [21], tree networks [22], Gaussian, mesh, torus networks [23], and complete b-Ary tree networks [24]. Generally, a load distribution strategy involves two main issues—one in deriving optimal sizes of the workloads to the processors, referred to as an optimal load partition (OLP), and the other is to determine a viable sequence of distribution that achieves minimum processing time, referred to as an optimal load distribution sequence (OLDS).

As for the first issue, in order to obtain a minimized processing time, it is necessary and sufficient to require that all processors stop computing at the same time instant; otherwise, the processing time of the entire workload could be reduced by transferring some load fractions from busy to idle processors. This widely accepted principle in Divisible-Load Theory (DLT) is referred to as the optimality principle, which provides a key to derive a closed-form solution for OLP [25]. However, as mentioned earlier, processors may have arbitrary unavailable time periods in reality. Hence, we could not inadvertently assign tasks to processors according to the optimality principle as usual; otherwise, workload rescheduling would result in an inefficient time and energy schedule. Therefore, searching for an OLP is necessary for sustainable computing where processor available time periods are involved.

As for the second issue, sufficient evidence has shown that load distribution sequences play a significant role in computational performance. For heterogeneous single-level tree networks, it has been proven that only when the load distribution sequence follows the decreasing order of communication speeds does the processing time reach the minimum [26]. As regard to heterogeneous multi-level tree networks, the OLDS depends only on communication speeds of links but not on computation speeds of processors [27]. Nonetheless, the above studies did not consider start-up overheads for both communication and computation into consideration. For the case of homogenous bus networks with start-up overheads, it was shown that the processing time is minimized when the load distribution sequence follows the order in which the computation speeds of processors decrease [28]. For a large enough workload on heterogeneous single-level tree networks with arbitrary start-up overheads, the sequence of load distribution should follow the decreasing order of the communication speeds in order to achieve minimum processing time [29], but how large a workload should be to consider it as a large enough workload. Moreover, when we consider processor available time periods, does the above conclusion still hold? If not, what sequence does the load distribution should follow to achieve a minimum processing time?

As regard to processor release times alone, several load distribution strategies were proposed for bus networks [30], linear daisy chain networks [31], and single-level tree networks [32], but they did not take start-up overheads and the influence of load distribution sequence into consideration. In order to obtain an OLP and OLDS simultaneously on single-level tree networks with arbitrary processor release times, a *bi-level genetic algorithm* was proposed in [33]. The proposed algorithm comprises two layers of nested genetic algorithms, with the upper genetic algorithm applied for searching an OLDS and the lower algorithm utilized for finding an OLP. However, as the number of processors increases, the proposed bi-level genetic algorithm gets hard to converge. In order to obtain an accurate OLP, an exhaustive search algorithm was proposed in [34] for release-time aware divisible-load scheduling on bus networks, but it did not consider processor off-line times and the influence of OLDS on processing time.

In this chapter, both processor release times and off-line times are explicitly considered in our model, which brings the work more closer to reality. This scheduling problem at hand is complex owing to an inherent nature of the computing platform which could possibly comprise heterogeneous processors. We propose a novel *Processor Availability-Aware Genetic Algorithm (PAA-GA)* based global optimization strategy to minimize the processing time of the entire workload, thus reducing the total energy consumption too, on HSCS.

The remaining of this chapter is organized as follows. Section 2 firstly gives a mathematical description of the divisible-load scheduling problem on HSCS with arbitrary start-up overheads and processor available time periods, followed by the proposed availability-aware scheduling model. With this model, we accordingly design algorithm PAA-GA in Sect. 3, which will be evaluated through experiments in Sect. 4. In the last section, conclusions are obtainable.

2 Availability-Aware Scheduling Model

2.1 Problem Description

An HSCS is considered in this chapter with its topology given in Fig. 1. It comprises $N + 1$ Heterogeneous processors $\{P_0, P_1, \ldots, P_N\}$ connected through communication links $\{L_1, L_2, \cdots, L_N\}$, where P_0 signifies the master, while the others

Fig. 1 An HSCS with $N + 1$ heterogeneous processors connected in a single-level tree topology

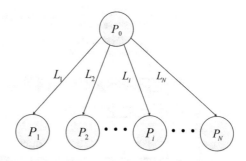

denote worker processors. P_0 does not participate in computation itself but merely takes the responsibility of assigning loads to worker processors. P_0 firstly divides the entire workload W_{total} into N fractions $\vec{A} = (\alpha_1, \alpha_2, \ldots, \alpha_N)$ with $0 \leq a_i \leq W_{total}$ and $\sum_{i=1}^{N} \alpha_i = W_{total}$. Then the load fractions are assigned to worker processors in a certain distribution order $(P_{\sigma_1}, P_{\sigma_2}, \ldots, P_{\sigma_N})$, where $\vec{\sigma} = (\sigma_1, \sigma_2, \ldots, \sigma_N)$ is processor index which is a permutation of $(1, 2, \ldots, N)$ and α_i is assigned to processor P_{σ_i} with $i = 1, 2, \ldots, N$. P_0 sends load fraction to only one processor at a time and each worker starts computing after its entire load fraction has been received completely. Workers cannot communicate and compute simultaneously.

It is necessary to note that, not all worker processors have necessity to participate in workload computation. Suppose that only the first n processors $P_{\sigma_1}, P_{\sigma_2}, \ldots, P_{\sigma_n}$ in the distribution sequence are needed for workload computation, so they will be assigned with non-zero load fractions, that is, $\alpha_i > 0$ with $i = 1, \ldots, n$, while the remaining processors are not assigned with any load fractions, that is, for $i = n + 1, \ldots, N$, $\alpha_i = 0$.

We consider a heterogeneous system wherein we have, for $\forall i \neq j$, $w_i \neq w_j$ and $g_i \neq g_j$. Also, it is assumed that communication speeds are much faster than computation speeds; otherwise, only one or two processors should be enough to involve in the workload computation [35]. As P_0 assigns α_i to the i-th processor P_{σ_i} in the load distribution sequence, the communication and computation components are modelled as affine functions, given by $e_{\sigma_i} + g_{\sigma_i}\alpha_i$ and $f_{\sigma_i} + w_{\sigma_i}\alpha_i$, including communication and computation start-up overheads e_{σ_i} and f_{σ_i} associated with processor P_{σ_i} and link L_{σ_i}, respectively.

Some processors in our system may be engaged in any of the previous workload computation when a new load arrives, say at time $t = 0$, so they cannot participate for the newly arrived workload computation until their release times. Meanwhile, they have to finish computing their assigned load fractions before they arrive at their off-line times. It is assumed that processors can estimate their release times by the size of the current workload to process, and the master knows the release and off-line times of all processors. Even though the master does not know the accurate processor off-line times, there exist some prediction techniques to estimate an approximate off-line time for each processor based on a history of processor usage (for more information, please refer to [36–38]). Let r_i and o_i be the release time and off-line time of processor P_i respectively, where $i = 1, 2, \ldots, N$.

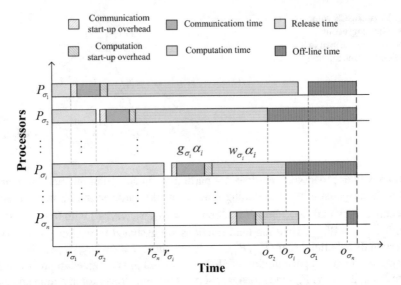

Fig. 2 Gantt chart for load scheduling on HSCS with processor available times

Figure 2 shows a possible Gantt Chart for load scheduling on HSCS with processor release times and off-line times. As illustrated in Fig. 2, the first processor P_{σ_1} starts to receive load fraction α_1 from master P_0 when it gets released at time $t = r_{\sigma_1}$. Let s_i be the start time of processor P_{σ_i}, at which P_{σ_i} starts to receive load fraction α_i from P_0. Thus $s_1 = r_{\sigma_1}$. Except for P_{σ_1}, the start time s_i of processor P_{σ_i} depends not only on its release time r_{σ_i}, but also on the start time s_{i-1} of processor $P_{\sigma_{i-1}}$ and the communication time $(e_{\sigma_{i-1}} + g_{\sigma_{i-1}} \alpha_{i-1})$ taken by processor $P_{\sigma_{i-1}}$ to receive its load fraction α_{i-1} from P_0. By observing Fig. 2, we obtain that $s_i = \max \left\{ r_{\sigma_i}, \, s_{i-1} + e_{\sigma_{i-1}} + g_{\sigma_{i-1}} \alpha_{i-1} \right\}$, where $i = 2, 3, \ldots, n$.

According to the optimality principle of DLT, if processor off-line times are ignored, all processors should finish computing at the same time to obtain a minimized processing time, say at time $t = T^\star$. Once we consider off-line times, processors whose off-line times are smaller than $T^\star$ will not be able to finish their assigned load fractions. Hence, lest these load fractions be rescheduled and result in a waste of energy consumption, processors should be assigned with appropriate workload sizes according to their available time periods in the first place.

The processing time T_i of processor P_{σ_i} is given by $s_i + e_{\sigma_i} + f_{\sigma_i} + \left(g_{\sigma_i} + w_{\sigma_i} \right) \alpha_i$, where $i = 1, 2, \ldots, n$. It can be observed from the formulation that T_i depends directly on two parts: the former part s_i indicating when processor P_{σ_i} starts to receive load from P_0, and the latter part $\left(e_{\sigma_i} + f_{\sigma_i} + \left(g_{\sigma_i} + w_{\sigma_i} \right) \alpha_i \right)$ representing how long it takes for processor P_{σ_i} to finish computing its assigned load fraction. Both parts are determined directly by load partition $\vec{A} = \{\alpha_1, \alpha_2, \ldots, \alpha_N\}$ and load distribution sequence $\left(P_{\sigma_1}, P_{\sigma_2}, \ldots, P_{\sigma_N} \right)$. Hence, T_i is actually a function of $\vec{\sigma}$ and $\vec{A}$. As the processing time T of the entire workload lies upon the processor which stops computing the last,

T can be derived as a function of $\vec{\sigma}$ and $\vec{A}$ as follows.

$$T(\vec{\sigma}, \vec{A}) = \max_{1 \le i \le n} T_i = \max_{1 \le i \le n} \left\{ s_i + e_{\sigma_i} + f_{\sigma_i} + \left(g_{\sigma_i} + w_{\sigma_i} \right) \alpha_i \right\}.$$

The objective of divisible-load scheduling on HSCS is to minimize the processing time T of the entire workload by taking into account processor available time periods, so that every amount of energy is consumed for useful workload computation without wasting, thus reducing the total energy consumption to the utmost. To achieve this goal, one has to determine an optimal load distribution strategy, including an OLDS and OLP. A feasible load distribution strategy should subject to the following four constraints:

(1) **Workload Constraint**: Each load fraction should be non-negative and not larger than the entire workload, the sum of which is equal to the entire workload. That is to say, $0 \le \alpha_i \le W_{total}$ with $i = 1, \ldots, N$, and $\sum_{i=1}^{N} \alpha_i = W_{total}$.

(2) **Processor Constraint**: A load distribution sequence should contain exactly one instance of a processor, without any omission or duplication of a processor or processors. That is, $\vec{\sigma} = (\sigma_1, \sigma_2, \ldots, \sigma_N)$, where $\sigma_i \in \{1, 2, \ldots, N\}$ with $i = 1, 2, \ldots, N$; for $\forall j, k \in \{1, 2, \ldots, N\}$, if $j \ne k$, then $\sigma_j \ne \sigma_k$.

(3) **Participant Constraint**: Not all processors are needed for workload computation. Assuming that only the first n ($n \le N$) processors in the distribution sequence are required, we have $\alpha_i > 0$ with $i = 1, 2, \ldots, n$, while $\alpha_i = 0$ when $i = n + 1, \ldots, N$.

(4) **Off-Line Time Constraint**: Processors involved in workload computation should stop computing before their off-line time come. That is, $T_i = s_i + e_{\sigma_i} + (g_{\sigma_i} + w_{\sigma_i})\alpha_i + f_{\sigma_i} \le o_{\sigma_i}$ with $i = 1, \ldots, n$.

2.2 A Novel Scheduling Model for Sustainable Computing

Table 1 briefly summarizes related notations and corresponding definitions. In order to solve the scheduling problem mentioned in the previous section, we build a novel processor availability-aware divisible-load scheduling model as follows:

$$\min_{\vec{\sigma}, \vec{A}} T(\vec{\sigma}, \vec{A}) = \min_{\vec{\sigma}, \vec{A}} \left\{ \max_{1 \le i \le n} \left\{ s_i + e_{\sigma_i} + f_{\sigma_i} + \left(g_{\sigma_i} + w_{\sigma_i} \right) \alpha_i \right\} \right\}.$$

s.t.

(1) $\sum_{i=1}^{N} \alpha_i = W_{total}$, $0 \le \alpha_i \le W_{total}$, $i = 1, \ldots, N$.

(2) $\vec{\sigma} = (\sigma_1, \sigma_2, \ldots, \sigma_N)$, where $\sigma_i \in \{1, 2, \ldots, N\}$ and $i = 1, 2, \ldots, N$. $\forall j, k \in \{1, 2, \ldots, N\}$, if $j \ne k$, then $\sigma_j \ne \sigma_k$.

(3) $n \le N$; $\forall i \in \{1, 2, \ldots, n\}$, $\alpha_i > 0$, while $\forall i \in \{n + 1, \ldots, N\}$, $\alpha_i = 0$.

(4) $s_i + e_{\sigma_i} + (g_{\sigma_i} + w_{\sigma_i}) \alpha_i + f_{\sigma_i} \le o_{\sigma_i}$, $i = 1, \ldots, n$.

Table 1 Notations and definitions

Notations	Definitions
W_{total}	Total size of the entire workload
N	Total number of worker processors
n	Number of processors required for workload computation
e_i	Communication start-up overhead of link L_i
f_i	Computation start-up overhead of processor P_i
g_i	Ratio of time taken by link L_i to communicate a given workload to that by a standard link
w_i	Ratio of time taken by processor P_i to compute a given workload to that by a standard processor
$\vec{\sigma}$	Processor index used for representing load distribution sequences. $\vec{\sigma} = (\sigma_1, \sigma_2, \ldots, \sigma_N)$ is a permutation of $(1, 2, \ldots, N)$
r_i	Release time of processor P_i
o_i	Off-line time of processor P_i
s_i	Start time of the i-th processor P_{σ_i} in the distribution sequence
$\vec{A}$	Load partition scheme. $\vec{A} = \{\alpha_1, \alpha_2, \ldots, \alpha_N\}$ with each element α_i representing the size of load fraction assigned to the i-th processor P_{σ_i} in the distribution sequence
T	Processing time of the entire workload
T_i	Processing time of the i-th processor P_{σ_i} in the distribution sequence

where

(5) $n = card(\{\alpha_i \mid \alpha_i \in \vec{A} \text{ and } \alpha_i > 0\})$, where $card(X)$ denotes the number of elements in set X.

(6) $s_i = \begin{cases} r_1, & i = 1; \\ \max\{r_{\sigma_i}, s_{i-1} + e_{\sigma_{i-1}} + g_{\sigma_{i-1}}\alpha_{i-1}\}, & i = 2, 3, \ldots, n. \end{cases}$

3 Algorithm PAA-GA Based Global Optimization Strategy

In the proposed model, two sets of variables are involved: $\vec{A} = \{\alpha_1, \alpha_2, \ldots, \alpha_N\}$ and $\vec{\sigma} = (\sigma_1, \sigma_2, \ldots, \sigma_N)$. Therefore, the solution of the proposed model is a mix of real numbers and integer numbers. The problem of deriving an OLDS is similar to Travelling Salesman Problem (TSP) which asks the following question: Given a list of cities and the distances between each pair of cities, what is the shortest possible route that visits each city exactly once and returns to the origin city? It is well acknowledged that TSP is an NP-hard problem in combinatorial optimization. Therefore, as an even more complex problem with two sets of variables $\vec{A}$ and $\vec{\sigma}$ optimized simultaneously, the problem considered in this chapter is definitely an NP-hard problem. When N turns out to be large, it is hard to obtain a global optimal solution $(\vec{A}, \vec{\sigma})$.

We select Genetic Algorithms (GAs), proposed by Holland [39], to solve our model because GAs have been proven to be a promising technique for task-scheduling problems, especially for complex permutation-based combinatorial optimization problems like TSP [40]. In this chapter, we shall first design an encoding scheme based on the characteristics of the proposed model, on the basis of which genetic operators are introduced, followed by the framework of PAA-GA.

3.1 Encoding Scheme

The key point of finding an optimal solution by using GAs is to develop an encoding scheme that can represent the problem to be solved directly and can satisfy the problem constraints easily. In this chapter, a hybrid encoding scheme is adopted. An individual is encoded as $\vec{I} = (\vec{\sigma}, \vec{A})$, where $\vec{\sigma} = (\sigma_1, \sigma_2, \ldots, \sigma_N)$ indicates processor index used for representing the load distribution sequence, and $\vec{A} = (\alpha_1, \alpha_2, \ldots, \alpha_N)$ stands for load partition scheme. If $\alpha_i = 0$, then it means processor P_{σ_i} does not participate in workload computation.

As a simple example, assume there are six worker processors and that the size of the entire workload is 1000 units. A possible encoding scheme is given as follows:

$$I = \begin{pmatrix} \vec{\sigma} \\ \vec{A} \end{pmatrix} = \begin{pmatrix} \sigma_1, \sigma_2, \sigma_3, \sigma_4, \sigma_5, \sigma_6 \\ \alpha_1, \alpha_2, \alpha_3, \alpha_4, \alpha_5, \alpha_6 \end{pmatrix} = \begin{pmatrix} 2, & 1, & 4, & 6, & 3, & 5 \\ 350, & 200, & 108, & 150, & 120, & 0 \end{pmatrix}.$$

$\vec{\sigma} = (2, 1, 4, 6, 3, 5)$ indicates that the load distribution sequence follows the order of $(P_2, P_1, P_4, P_6, P_3, P_5)$ and $\vec{A} = (350, 200, 180, 150, 120, 0)$ means that the sizes of load fractions assigned to processors $P_2, P_1, P_4, P_6, P_3, P_5$ are 350, 200, 180, 150, 120, 0, respectively. Note that, only 5 processors take part in workload computation since $\alpha_6 = 0$, which means P_0 does not assign any load to the last processor P_5 in the distribution sequence, so P_5 does not participate in workload computation. Therefore, we have $n = 5$.

3.2 Crossover Operators

It is worth noting that given a deterministic load distribution sequence, there exists an OLP that achieves minimum processing time of the entire workload. Hence, the scheduling problem dealt in this chapter can be summarized as determining an OLDS, on the basis of which deriving an OLP to achieve minimum processing time. Therefore, the scheduling problem has actually two layers of decision levels, with the upper level determining an OLDS and the lower level deriving an OLP. With this consideration in mind, we design two crossover operators in this section. One is to optimize the upper level variable $\vec{\sigma}$ as well as the lower level variable $\vec{A}$ simultaneously, while the other is to optimize $\vec{A}$ alone based on a fixed $\vec{\sigma}$.

Also noteworthy is the fact that both $\vec{\sigma}$ and $\vec{A}$ are not suitable for traditional two-point crossover. This is because $\vec{\sigma}$ must contain exactly one instance of a number in $\{1, 2, \ldots, N\}$ and any omission or duplication of numbers leads to an invalid solution. Meanwhile, $\vec{A}$ has to satisfy that $\sum_{i=1}^{N} \alpha_i = W_{total}$.

3.2.1 First Crossover Operator

The first way of crossover is to evolve $\vec{\sigma}$ and $\vec{A}$ at the same time. For two parents $\vec{I}^1 = \left(\vec{\sigma}^1, \vec{A}^1\right)$ and $\vec{I}^2 = \left(\vec{\sigma}^2, \vec{A}^2\right)$, the steps given in Algorithm 1 are adopted to generate two offsprings $\vec{I}^3 = \left(\vec{\sigma}^3, \vec{A}^3\right)$ and $\vec{I}^4 = \left(\vec{\sigma}^4, \vec{A}^4\right)$.

Algorithm 1 First Crossover Operator

Input: Two parents $\vec{I}^1 = \left(\vec{\sigma}^1, \vec{A}^1\right)$ and $\vec{I}^2 = \left(\vec{\sigma}^2, \vec{A}^2\right)$.

Output: Two offsprings $\vec{I}^3 = \left(\vec{\sigma}^3, \vec{A}^3\right)$ and $\vec{I}^4 = \left(\vec{\sigma}^4, \vec{A}^4\right)$.

1: Generate two random integers p and q between 1 and N as the crossover points that satisfy $1 \le p < q \le N$.

2: **for** $i = p, \ldots, q$ **do**

3:     exchange genes σ_i^1 and σ_i^2 to obtain σ_i^3 and σ_i^4. Let $\sigma_i^3 = \sigma_i^2$ and $\sigma_i^4 = \sigma_i^1$.

4:     exchange genes α_i^1 and α_i^2 to obtain α_i^3 and α_i^4. Let $\alpha_i^3 = \alpha_i^2$ and $\alpha_i^4 = \alpha_i^1$.

5: **end for**

6: **for** $i = 1, \ldots, p-1$ and $i = q+1, \ldots, N$ **do**

7:     let $\sigma_i^3 = \sigma_i^1, \sigma_i^4 = \sigma_i^2, \alpha_i^3 = \alpha_i^1$, and $\alpha_i^4 = \alpha_i^2$.

8: **end for**

9: Establish mapping relationships for interchangeability based on the genes of $\vec{\sigma}^1$ and $\vec{\sigma}^2$ between the two crossover points.

10: Based on the mapping relationships, replace the genes of $\vec{\sigma}^3$ and $\vec{\sigma}^4$ outside the two crossover points that have the same value with genes inside.

11: Normalize $\vec{A}^3$ and $\vec{A}^4$ to ensure that the total size of all load fractions equals the entire workload W_{total}.

For instance, we have the following two parents.

$$\vec{I}^1 = \begin{pmatrix} \vec{\sigma}^1 \\ \vec{A}^1 \end{pmatrix} = \begin{pmatrix} 2, & | & 1, & 4, & 6, & | & 3, & 5 \\ 350, & | & 200, & 180, & 150, & | & 120, & 0 \end{pmatrix}.$$

$$\vec{I}^2 = \begin{pmatrix} \vec{\sigma}^2 \\ \vec{A}^2 \end{pmatrix} = \begin{pmatrix} 4, & | & 3, & 6, & 2, & | & 5, & 1 \\ 320, & | & 270, & 160, & 140, & | & 150, & 60 \end{pmatrix}.$$

According to Step 1 of Algorithm 1, we first generate two random integers p and q. Suppose that $p = 2$ and $q = 4$. After exchanging of genes based on Steps 2–8, we then have,

$$\vec{I}^3 = \begin{pmatrix} \vec{\sigma}^3 \\ \vec{A}^3 \end{pmatrix} = \begin{pmatrix} 2, & | & 3, & 6, & 2, & | & 3, & 5 \\ 350, & | & 270, & 160, & 140, & | & 120, & 0 \end{pmatrix}.$$

$$\vec{I}^4 = \begin{pmatrix} \vec{\sigma}^4 \\ \vec{A}^4 \end{pmatrix} = \begin{pmatrix} 4, & | & 1, & 4, & 6, & | & 5, & 1 \\ 320, & | & 200, & 180, & 150, & | & 150, & 60 \end{pmatrix}.$$

We observe that both $\vec{I}^3$ and $\vec{I}^4$ are invalid solutions because some genes of $\vec{\sigma}^3$ and $\vec{\sigma}^4$ that are outside the two crossover points have the same values as the genes between the two points, and also that the total size of load fractions is not equal to the entire workload. Therefore, we need to adjust $\vec{I}^3$ and $\vec{I}^4$.

According to Step 9 of Algorithm 1, we establish the following mapping relationships,

$$1 \Leftrightarrow 3, \quad 4 \Leftrightarrow 6, \text{ and } 6 \Leftrightarrow 2 \quad \text{implies} \quad 1 \Leftrightarrow 3 \text{ and } 4 \Leftrightarrow 2.$$

With the above mapping relationships, we can fix $\vec{\sigma}^3$ and $\vec{\sigma}^4$ now. As for $\vec{\sigma}^3$, $\sigma_1^3 = 2$ should be replaced by 4, $\sigma_5^3 = 3$ replaced by 1, and $\sigma_6^3 = 5$ remaining unchanged. Similarly, as for $\vec{\sigma}^4$, $\sigma_1^4 = 4$ should be replaced by 2, $\sigma_5^4 = 5$ remaining unchanged, and $\sigma_6^4 = 1$ replaced by 3. Hence, we have,

$$\vec{I}^3 = \begin{pmatrix} \vec{\sigma}^3 \\ \vec{A}^3 \end{pmatrix} = \begin{pmatrix} 4, & | & 3, & 6, & 2, & | & 1, & 5 \\ 350, & | & 270, & 160, & 140, & | & 120, & 0 \end{pmatrix}.$$

$$\vec{I}^4 = \begin{pmatrix} \vec{\sigma}^4 \\ \vec{A}^4 \end{pmatrix} = \begin{pmatrix} 2, & | & 1, & 4, & 6, & | & 5, & 3 \\ 320, & | & 200, & 180, & 150, & | & 150, & 60 \end{pmatrix}.$$

After normalization by the last step of Algorithm 1, we obtain two offsprings as follows.

$$\vec{I}^3 = \begin{pmatrix} \vec{\sigma}^3 \\ \vec{A}^3 \end{pmatrix} = \begin{pmatrix} 4, & | & 3, & 6, & 2, & | & 1, & 5 \\ 317, & | & 267, & 158, & 139, & | & 119, & 0 \end{pmatrix}.$$

$$\vec{I}^4 = \begin{pmatrix} \vec{\sigma}^4 \\ \vec{A}^4 \end{pmatrix} = \begin{pmatrix} 2, & | & 1, & 4, & 6, & | & 5, & 3 \\ 321, & | & 183, & 165, & 138, & | & 138, & 55 \end{pmatrix}.$$

3.2.2 Second Crossover Operator

The second way of crossover is to evolve $\vec{A}$ based on a fixed $\vec{\sigma}$. Algorithm 2 shows its main steps. As an example, suppose $p = 2$ and $q = 5$. By Steps 2–5 of Algorithm 2, we have,

$$\vec{I}^1 = \begin{pmatrix} \vec{\sigma}^1 \\ \vec{A}^1 \end{pmatrix} = \begin{pmatrix} 2, & | & 1, & 4, & 6, & 3, & | & 5 \\ 350, & | & 200, & 180, & 150, & 120, & | & 0 \end{pmatrix}.$$

$$\vec{I}^2 = \begin{pmatrix} \vec{\sigma}^2 \\ \vec{A}^2 \end{pmatrix} = \begin{pmatrix} 4, & | & 3, & 6, & 2, & 5, & | & 1 \\ 320, & | & 270, & 160, & 140, & 150, & | & 60 \end{pmatrix}.$$

$$\vec{I}^3 = \begin{pmatrix} \vec{\sigma}^3 \\ \vec{A}^3 \end{pmatrix} = \begin{pmatrix} 2, & | & 1, & 4, & 6, & 3, & | & 5 \\ 350, & | & 270, & 160, & 140, & 150, & | & 0 \end{pmatrix}.$$

$$\vec{I}^4 = \begin{pmatrix} \vec{\sigma}^4 \\ \vec{A}^4 \end{pmatrix} = \begin{pmatrix} 4, & | & 3, & 6, & 2, & 5, & | & 1 \\ 320, & | & 200, & 180, & 150, & 120, & | & 60 \end{pmatrix}.$$

It is worth noting that both $\vec{I}^3$ and $\vec{I}^4$ are invalid solutions because $\sum_{i=1}^{N} \alpha_i^3 < W_{total}$ and $\sum_{i=1}^{N} \alpha_i^4 > W_{total}$. After normalization by Step 6, we obtain two offsprings as follows.

$$\vec{I}^3 = \begin{pmatrix} \vec{\sigma}^3 \\ \vec{A}^3 \end{pmatrix} = \begin{pmatrix} 2, & | & 1, & 4, & 6, & 3, & | & 5 \\ 327, & | & 252, & 150, & 131, & 140, & | & 0 \end{pmatrix}.$$

$$\vec{I}^4 = \begin{pmatrix} \vec{\sigma}^4 \\ \vec{A}^4 \end{pmatrix} = \begin{pmatrix} 4, & | & 3, & 6, & 2, & 5, & | & 1 \\ 311, & | & 194, & 185, & 146, & 116, & | & 58 \end{pmatrix}.$$

Algorithm 2 Second Crossover Operator

Input: Two parents $\vec{I}^1 = \left(\vec{\sigma}^1, \vec{A}^1 \right)$ and $\vec{I}^2 = \left(\vec{\sigma}^2, \vec{A}^2 \right)$.

Output: Two offsprings $\vec{I}^3 = \left(\vec{\sigma}^3, \vec{A}^3 \right)$ and $\vec{I}^4 = \left(\vec{\sigma}^4, \vec{A}^4 \right)$.

1: Let $\vec{I}^3 = \vec{I}^1$ and $\vec{I}^4 = \vec{I}^2$.
2: Randomly select two crossover points p and q that satisfy $1 \leq p < q \leq N$.
3: **for** $i = p, \ldots, q$ **do**
4:     exchange genes α_i^3 and α_i^4. Let $\alpha_i^3 = \alpha_i^2$ and $\alpha_i^4 = \alpha_i^1$.
5: **end for**
6: Normalize $\vec{A}^3$ and $\vec{A}^4$ to ensure that the total size of all load fractions is equal to that of the entire workload.

3.3 Mutation Operator

The purpose of mutation in GAs focuses on preserving and introducing diversity from one generation of a population to the next. It is analogous to biological mutation. Mutation operator alters one or more gene values in an individual from its initial state, thus avoiding local minima and preventing the population from becoming too similar to each other. Mutation occurs to offsprings generated by crossover according to a user-definable mutation probability. This probability should be set low; otherwise, the search will turn into a primitive random search [41].

We apply two-point mutation on both $\vec{\sigma}$ and $\vec{A}$ simultaneously to obtain a new offspring. Randomly generate four integers p, q, l, and m that satisfy $1 \leq p < q \leq N$ and $1 \leq l < m \leq N$. Exchange genes σ_p and σ_q, as well as α_l and α_m. For example, suppose $p = 2$, $q = 4$, $l = 1$, and $m = 5$. We obtain offspring $\vec{I}'$ mutated from $\vec{I}$ as follows.

$$\vec{I} = \left(\begin{array}{c} \vec{\sigma} \\ \vec{A} \end{array} \right) = \left(\begin{array}{cccccc} 2, & \mathbf{1}, & 4, & \mathbf{6}, & 3, & 5 \\ \mathbf{350}, & 200, & 180, & 150, & \mathbf{120}, & 0 \end{array} \right).$$

$$\vec{I}' = \left(\begin{array}{c} \vec{\sigma}' \\ \vec{A}' \end{array} \right) = \left(\begin{array}{cccccc} 2, & \mathbf{6}, & 4, & \mathbf{1}, & 3, & 5 \\ \mathbf{120}, & 200, & 180, & 150, & \mathbf{350}, & 0 \end{array} \right).$$

3.4 Repair Operator

It cannot be expected that crossover and mutation operators produce new offsprings that satisfy all of the four constraints in our proposed model by default, especially for the Participant Constraint and Off-Line Time Constraint. Thus a newly generated individual need to be checked whether it violates either of the constraints. If so, we have to repair it to a feasible solution. Algorithm 3 gives the main steps of repair operator with the first two steps responsible for Participant Constraint satisfaction and the remaining for Off-Line Time Constraint satisfaction.

Figure 3 shows a Gantt Chart with processors that violate the Off-Line Time Constraint of the proposed model. It can be observed from Fig. 3 that the processing time T_2 of processor P_{σ_2} exceeds its off-line time o_{σ_2}, which results in a time conflict. Therefore, the individual that corresponds to this Gantt Chart violates the Off-Line Time Constraint and needs to be repaired. According to the repair operator, excessive load assigned to P_{σ_2} that causes the time conflict should be removed to its immediate successor P_{σ_3}. Fig. 4 shows a possible Gantt Chart after the load distribution adjustment. It can be seen that $T_2 = o_{\sigma_2}$, so the time conflict for processor P_{σ_2} has been eliminated. However, we also notice that the load adjustment gives rise to a new time conflict for processor P_{σ_3}, whose processing time T_3 exceeds its off-line time o_{σ_3}. Therefore, another round of adjustment is required for processor P_{σ_3}. According to the repair operator, excessive load assigned to P_{σ_3} will be scheduled to its immediate successor P_{σ_4}. This process repeats iteratively until there are no time conflicts for all processors.

One may notice that we keep reallocating excessive load from a processor to its successors in the distribution sequence, instead of its predecessors. This is because

---

Algorithm 3 Repair Operator

1: For a newly generated individual $\vec{I} = (\vec{\sigma}, \vec{A})$, calculate the number of processors participating in workload computation by $n = card(\{\alpha_i \mid \alpha_i \in \vec{A} \text{ and } \alpha_i > 0\})$.
   For an instance, given that

$$\vec{I} = \begin{pmatrix} \vec{\sigma} \\ \vec{A} \end{pmatrix} = \begin{pmatrix} \sigma_1, & \sigma_2, & \sigma_3, & \sigma_4, & \sigma_5, & \sigma_6 \\ \alpha_1, & \alpha_2, & \alpha_3, & \alpha_4, & \alpha_5, & \alpha_6 \end{pmatrix} = \begin{pmatrix} 1, & 2, & 3, & 4, & 5, & 6 \\ 327, & 0, & 0, & 182, & 140, & 151 \end{pmatrix}.$$

   We have $n = card(\{327, 182, 140, 151\}) = 4$.
2: Adjust the order of (σ_i, α_i) pairs so that processors with non-zero load fractions are listed in front of the others.
   As for the above example, it can be observed that $\alpha_2 = \alpha_3 = 0$, so pairs (σ_2, α_2) and (σ_3, α_3) should be placed at the end of $\vec{I}$. After reordering, we obtain,

$$\vec{I} = \begin{pmatrix} \vec{\sigma} \\ \vec{A} \end{pmatrix} = \begin{pmatrix} 1, & 4, & 5, & 6, & 2, & 3 \\ 327, & 182, & 140, & 151, & 0, & 0 \end{pmatrix}.$$

3: Let $s_1 = r_{\sigma_1}$.
4: **for** $i = 2, 3, \ldots, n$ **do**
5:     calculate the start time of each processor by $s_i = \max\{r_{\sigma_i}, s_{i-1} + e_{\sigma_{i-1}} + g_{\sigma_{i-1}} \alpha_{i-1}\}$.
6: **end for**
7: **for** $i = 1, 2, \ldots, n$ **do**
8:     compute the processing time of each processor by $T_i = s_i + (g_{\sigma_i} + w_{\sigma_i}) \alpha_i + e_{\sigma_i} + f_{\sigma_i}$.
9: **end for**
10: **for** $i = 1, 2, \ldots, n$ **do**
11:     check whether processing time T_i of processor P_{σ_i} exceeds its off-line time o_{σ_i}. If $T_i \leq o_{\sigma_i}$ holds for all processors, then stop; otherwise, $T_i > o_{\sigma_i}$ means too much load has been assigned to processor P_{σ_i} and that excessive load needs to be rescheduled.
12: **end for**
13: Compute the size of excess load by $\Delta = (T_i - o_{\sigma i}) / (g_{\sigma i} + w_{\sigma i})$.
14: Assign this excess load fraction Δ to the next processor $P_{\sigma_{i+1}}$ by $\alpha_{i+1} = \alpha_{i+1} + \Delta$ and $\alpha_i = \alpha_i - \Delta$. Then go back to Step 1.

---

the start time of each processor is determined by the load fractions assigned to its predecessors according to the equation $s_i = \max\{r_{\sigma_i}, s_{i-1} + e_{\sigma_{i-1}} + g_{\sigma_{i-1}} \alpha_{i-1}\}$. Suppose processor P_i violates the Off-Line Time Constraint. If we reallocate excessive load from P_i to one of the processors before P_i in the distribution sequence, say P_j, then the start times of processors behind P_j may all get postponed, which has a high chance of causing more time conflicts for processors between P_i and P_j, thus taking a much longer time for repair operator to fix all time conflicts.

3.5 Local Search

As mentioned earlier, searching for an OLDS itself is already an NP-hard problem, not to mention that we also have to search for an OLP. When an HSCS scales up to a large number of processors, it may hard for GAs to converge. In order to improve convergence speed of the proposed GA, we introduce a local search operator. The main idea is to balance load between processors with the longest processing time T_{max} and the shortest processing time T_{min}, so that all of the processors with processing times not up to their off-line times will eventually stop computing at the same time. The process of the local search operator is given in Algorithm 4.

Algorithm 4 Local Search

1: For a given individual $\vec{I} = (\vec{\sigma}, \vec{A})$, calculate the number n of processors participating in the workload computation by $n = card(\{\alpha_i \mid \alpha_i \in \vec{A} \ and \ \alpha_i > 0\})$.
2: Let $s_1 = r_{\sigma_1}$.
3: **for** $i = 2, 3, \ldots, n$ **do**
4:     calculate the start time of each processor by $s_i = \max\{r_{\sigma_i}, \ s_{i-1} + e_{\sigma_{i-1}} + g_{\sigma_{i-1}} \alpha_{i-1}\}$.
5: **end for**
6: **for** $i = 1, 2, \ldots, n$ **do**
7:     compute the processing time of each processor by $T_i = s_{\sigma_i} + (g_{\sigma_i} + w_{\sigma_i}) \alpha_i + e_{\sigma_i} + f_{\sigma_i}$.
8: **end for**
9: Among $P_{\sigma_1}, P_{\sigma_2}, \ldots, P_{\sigma_n}$, find processor $P_{\sigma_{max}}$ with the longest processing time T_{max} and processor $P_{\sigma_{min}}$ with the shortest processing time T_{min}. Calculate their time difference by $\Delta = (T_{max} - T_{min})$.
10: Let $\beta = (T_{max} - T_{min}) / \max\{g_{\sigma_{max}}, g_{\sigma_{min}}\}$. Update individual $\vec{I} = (\vec{\sigma}, \vec{A})$ by $\alpha_{max} = \alpha_{max} - \beta$ and $\alpha_{min} = \alpha_{min} + \beta$.
11: Apply repair operator given by Algorithm 3 on the updated individual to ensure that it satisfies all constraints of the proposed model.

Figure 5 shows a Gantt Chart that corresponds to an individual before applying local search operator. As illustrated in Fig. 5, processor P_{σ_1} has the longest processing

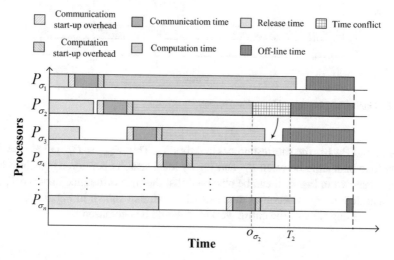

Fig. 3 Gantt Chart with processors violating the Off-Line Time Constraint of the proposed model

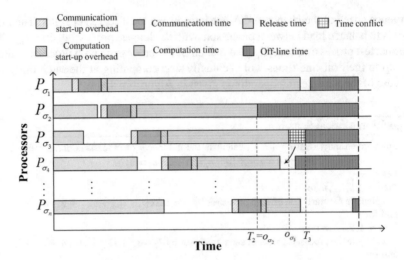

Fig. 4 Gantt Chart after load distribution adjustment by repair operator

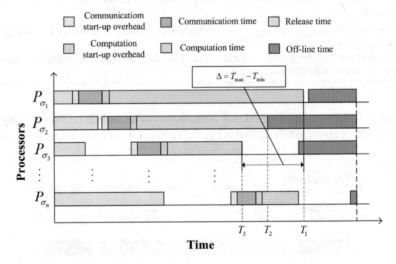

Fig. 5 Gantt Chart before applying local search

time T_1 and P_{σ_3} has the shortest processing time T_3. Thus $P_{\max} = P_{\sigma_1}$ and $P_{\min} = P_{\sigma_3}$. After load balancing between P_{σ_1} and P_{σ_3} by local search operator, a possible Gantt Chart is shown in Fig. 6. It can be observed that the time difference between T_1 and T_3 illustrated in Fig. 6 becomes much smaller than that shown in Fig. 5. Hence, the total processing time of the entire workload would be decreased.

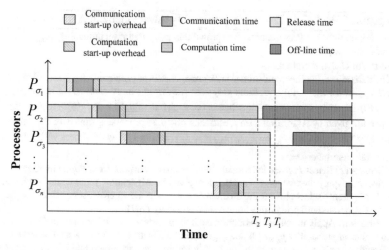

Fig. 6 Gantt Chart after applying local search

3.6 Framework of Algorithm PAA-GA

Once encoding scheme is defined, a GA initializes a population of individuals and then improves them through repetitive applications of genetic operators, including crossover, mutation, repair, local search, and selection. The framework and flow chart of algorithm PAA-GA are given in Algorithm 5 and Fig. 7, respectively.

4 Experimental Results and Analysis

Several rigours experiments are conducted to study the performance and demonstrate the effectiveness of the proposed algorithm. We employed a compute cluster comprising 15 nodes and the parameters of our HSCS are given in Table 2. In the master node where our proposed scheduling algorithm PAA-GA runs, the following parameters are set: population size $Popsize = 100$, crossover probability $p_{cros} = 0.6$, mutation probability $p_{mut} = 0.02$, elitist number $E = 5$, and stop criterion $t = 50000$.

4.1 Evaluating the Correctness of PAA-GA

We attempt to make a comparison between an exhaustive algorithm (EA) [34] with our PAA-GA. As expected, although EA will be time-consuming, EA can obtain an absolute minimum processing time. In order to evaluate the correctness of algorithm PAA-GA, we make a comparison between PAA-GA and EA to check whether

---

Algorithm 5 PAA-GA: Processor Availability-Aware Genetic Algorithm

---

Input: Population size *Popsize*, crossover probability p_{cros}, mutation probability p_{mut}, elitist number E and stop criterion.

Output: An OLDS $\vec{\sigma}$ and OLP $\vec{A}$.

1: (**Initialization**) Set the population size *Popsize*, crossover probability p_{cros}, mutation probability p_{mut}, and elitist number E. Randomly generate *Popsize* individuals as the initial population $Pop(0)$ according to the encoding scheme. For each individual $\vec{I} \in Pop(0)$, first apply the repair operator given by Algorithm 3 on $\vec{I}$, and then compute processing time T of the entire workload by $T = \max_{1 \le i \le n} \left\{ s_i + e_{\sigma_i} + \left(g_{\sigma_i} + w_{\sigma_i} \right) \alpha_i + f_{\sigma_i} \right\}$, taking $1/T$ as the fitness value of $\vec{I}$. Let the generation number $t = 0$.

2: (**Crossover**) Select *Popsize* individuals into the crossover pool from $Pop(t)$ by roulette wheel selection. Apply the two crossover operators given by Algorithms 1 and 2 one-by-one on each pair of parents selected from the crossover pool according to crossover probability p_{cros}. All newly generated offsprings constitute a set denoted by $O_1(t)$.

3: (**Mutation**) Apply mutation operator on each of the selected individuals from $O_1(t)$ according to mutation probability p_{mut}. All newly generated offsprings constitute a set denoted by $O_2(t)$.

4: (**Repair**) Apply repair operator given by Algorithm 3 on each individual in set $O_1(t) \cup O_2(t)$.

5: (**Local Search**) Apply local search operator given by Algorithm 4 on each individual in set $O_1(t) \cup O_2(t)$.

6: (**Selection**) Select the best E individuals for the next population $Pop(t + 1)$ from set $Pop(t) \cup O_1(t) \cup O_2(t)$. Select the remaining $Popsize - E$ individuals for $Pop(t + 1)$ by roulette wheel selection also from set $Pop(t) \cup O_1(t) \cup O_2(t)$. Let $t = t + 1$.

7: (**Stopping Criteria**) If a fixed number of generations reached, then stop and return the best individual $\vec{I} = (\vec{\sigma}, \vec{A})$ in the current population; otherwise, go to Step 2.

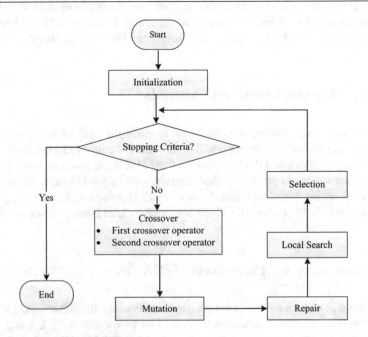

Fig. 7 Flow chart of algorithm PAA-GA

Table 2 Parameters of our HSCS

P_i	g_i	w_i	e_i	f_i	r_i	o_i
P_1	0.53	2.90	7.06	5.80	46.34	331.85
P_2	0.77	7.61	3.02	0.14	53.55	397.44
P_3	0.71	4.14	8.14	0.45	78.14	420.58
P_4	0.79	9.62	8.63	3.74	93.47	509.31
P_5	0.06	3.64	8.71	9.50	13.86	613.82
P_6	0.77	5.92	5.25	0.54	57.31	722.11
P_7	0.30	6.48	4.69	6.23	47.47	855.85
P_8	0.28	8.25	2.64	8.30	22.18	964.82
P_9	0.99	2.27	5.89	9.11	34.38	1175.66
P_{10}	0.98	5.34	6.95	2.44	17.06	1299.34
P_{11}	0.99	1.57	1.06	6.76	79.63	1474.28
P_{12}	0.10	7.99	5.75	1.03	31.81	1763.75
P_{13}	0.05	3.82	2.84	2.96	53.32	1768.40
P_{14}	0.95	4.01	3.01	9.80	17.40	1911.18
P_{15}	0.16	6.47	2.78	1.63	60.60	1943.74

the processing time obtained by PAA-GA agrees with that obtained by EA. If their processing times are in good agreement, then it surely proves that PAA-GA can obtain an OLP.

Assuming that the off-line times of all processors are infinite, thus we only take the processor release times into account. Note that EA requires a fixed load distribution sequence as its input in advance, so we set the OLDS obtained by PAA-GA as the input for algorithm EA. Table 3 records the comparison results for PAA-GA and EA. It can be observed from this table that PAA-GA obtains the same experimental results with EA for each test workload, including the same number of processors involved in workload computation and the same processing time. Therefore, we can make the conclusion that algorithm PAA-GA proposed in this chapter can obtain an OLP such that the processing time is minimized for divisible-load scheduling problems with processor release times.

Besides OLP, in order to prove that the proposed PAA-GA can also obtain an OLDS, we make a comparison between PAA-GA and EA with three different distribution sequences as its input, which are commonly used in previous studies: sequence in the order of increasing value of g_i, denoted as IG; sequence in the order of increasing value of w_i, denoted as IW; and sequence in the order of increasing value of release time r_i, denoted as IR.

Based on the parameters given in Table 2, sequences IG, IW, and IR are as follows,

Table 3 Experimental results obtained by PAA-GA and EA with the same OLDS

W_{total}	Algorithm	n	T	$\vec{\sigma} = (\sigma_1, \sigma_2, \ldots, \sigma_n)$
100	PAA-GA	12	110.615	(5, 8, 14, 12, 7, 13, 15, 1, 2, 11, 6, 3)
	EA	12	110.615	(5, 8, 14, 12, 7, 13, 15, 1, 2, 11, 6, 3)
200	PAA-GA	13	165.639	(5, 8, 12, 14, 13, 15, 7, 1, 11, 9, 2, 6, 3)
	EA	13	165.639	(5, 8, 12, 14, 13, 15, 7, 1, 11, 9, 2, 6, 3)
300	PAA-GA	14	216.182	(5, 8, 12, 7, 13, 15, 1, 3, 2, 11, 14, 9, 6, 10)
	EA	14	216.182	(5, 8, 12, 7, 13, 15, 1, 3, 2, 11, 14, 9, 6, 10)
400	PAA-GA	15	265.534	(5, 8, 12, 7, 13, 15, 1, 3, 2, 6, 11, 14, 9, 10, 4)
	EA	15	265.534	(5, 8, 12, 7, 13, 15, 1, 3, 2, 6, 11, 14, 9, 10, 4)
500	PAA-GA	15	314.943	(5, 8, 12, 7, 13, 15, 1, 3, 2, 6, 14, 11, 9, 10, 4)
	EA	15	314.943	(5, 8, 12, 7, 13, 15, 1, 3, 2, 6, 14, 11, 9, 10, 4)
600	PAA-GA	15	365.878	(5, 8, 12, 13, 15, 7, 1, 3, 2, 6, 14, 11, 9, 10, 4)
	EA	15	365.878	(5, 8, 12, 13, 15, 7, 1, 3, 2, 6, 14, 11, 9, 10, 4)
700	PAA-GA	15	414.793	(5, 8, 12, 13, 15, 7, 1, 3, 2, 6, 14, 11, 9, 10, 4)
	EA	15	414.793	(5, 8, 12, 13, 15, 7, 1, 3, 2, 6, 14, 11, 9, 10, 4)
800	PAA-GA	15	465.835	(5, 8, 12, 13, 15, 7, 1, 3, 2, 6, 14, 11, 9, 10, 4)
	EA	15	465.835	(5, 8, 12, 13, 15, 7, 1, 3, 2, 6, 14, 11, 9, 10, 4)
900	PAA-GA	15	516.877	(5, 8, 12, 13, 15, 7, 1, 3, 2, 6, 14, 11, 9, 10, 4)
	EA	15	516.877	(5, 8, 12, 13, 15, 7, 1, 3, 2, 6, 14, 11, 9, 10, 4)
1000	PAA-GA	15	567.761	(5, 12, 8, 13, 15, 7, 1, 3, 2, 6, 14, 11, 9, 10, 4)
	EA	15	567.761	(5, 12, 8, 13, 15, 7, 1, 3, 2, 6, 14, 11, 9, 10, 4)

$$IG = (13, 5, 12, 15, 8, 7, 1, 3, 2, 6, 4, 15, 10, 9, 11).$$
$$IW = (11, 9, 1, 5, 13, 14, 3, 10, 6, 15, 7, 2, 12, 8, 4).$$
$$IR = (5, 10, 14, 8, 12, 9, 1, 7, 13, 2, 6, 15, 3, 11, 4).$$

Table 4 records the experimental results obtained by algorithms PAA-GA, EA-IG, EA-IW, and EA-IR. It can be observed from Table 4 that for each test workload, the load distribution sequence obtained by PAA-GA is different from IG, IW, and IR. Moreover, for some test workloads, PAA-GA even obtains different numbers of processors involved in workload computation from EA with IG, IW, and IR, hence obtaining different load partition too. To be more intuitively, Fig. 8 illustrates the variation of processing time obtained by PAA-GA, EA-IG, EA-IW, and EA-IR along with different workload size. From this figure, we can observe that for each workload, the processing time obtained by the proposed PAA-GA is less than that by EA with three different load distribution sequences. The processing time obtained by PAA-GA shows a gain of about 10–25% compared to EA with IG and IR, and gained over 40% compared to EA with IW. Therefore, it is clear that PAA-GA outperforms over other strategies in achieving an optimal processing time, hence an efficient energy consumption too, for processor release time-aware divisible-load scheduling. Fur-

Table 4 Experimental results obtained by PAA-GA and EA with three different load distribution sequences

W_{total}	Algorithm	n	T	$\vec{\sigma} = (\sigma_1, \sigma_2, \ldots, \sigma_n)$
100	PAA-GA	12	110.615	(5, 8, 14, 12, 7, 13, 15, 1, 2, 11, 6, 3)
	EA-IG	11	145.418	(13, 5, 12, 15, 8, 7, 1, 3, 2, 6, 4)
	EA-IW	8	208.662	(11, 9, 1, 5, 13, 14, 3, 10)
	EA-IR	12	130.256	(5, 10, 14, 8, 12, 9, 1, 7, 13, 2, 6, 15)
200	PAA-GA	13	165.639	(5, 8, 12, 14, 13, 15, 7, 1, 11, 9, 2, 6, 3)
	EA-IG	15	201.189	(13, 5, 12, 15, 8, 7, 1, 3, 2, 6, 4, 14, 10, 9, 11)
	EA-IW	12	297.881	(11, 9, 1, 5, 13, 14, 3, 10, 6, 15, 7, 2)
	EA-IR	14	199.450	(5, 10, 14, 8, 12, 9, 1, 7, 13, 2, 6, 15, 3, 11)
300	PAA-GA	14	216.182	(5, 8, 12, 7, 13, 15, 1, 3, 2, 11, 14, 9, 6, 10)
	EA-IG	15	251.701	(13, 5, 12, 15, 8, 7, 1, 3, 2, 6, 4, 14, 10, 9, 11)
	EA-IW	14	382.499	(11, 9, 1, 5, 13, 14, 3, 10, 6, 15, 7, 2, 12, 8)
	EA-IR	15	265.008	(5, 10, 14, 8, 12, 9, 1, 7, 13, 2, 6, 15, 3, 11, 4)
400	PAA-GA	15	265.534	(5, 8, 12, 7, 13, 15, 1, 3, 2, 6, 11, 14, 9, 10, 4)
	EA-IG	15	302.213	(13, 5, 12, 15, 8, 7, 1, 3, 2, 6, 4, 14, 10, 9, 11)
	EA-IW	15	465.058	(11, 9, 1, 5, 13, 14, 3, 10, 6, 15, 7, 2, 12, 8, 4)
	EA-IR	15	330.451	(5, 10, 14, 8, 12, 9, 1, 7, 13, 2, 6, 15, 3, 11, 4)
500	PAA-GA	15	314.943	(5, 8, 12, 7, 13, 15, 1, 3, 2, 6, 14, 11, 9, 10, 4)
	EA-IG	15	352.725	(13, 5, 12, 15, 8, 7, 1, 3, 2, 6, 4, 14, 10, 9, 11)
	EA-IW	15	547.376	(11, 9, 1, 5, 13, 14, 3, 10, 6, 15, 7, 2, 12, 8, 4)
	EA-IR	15	395.894	(5, 10, 14, 8, 12, 9, 1, 7, 13, 2, 6, 15, 3, 11, 4)

thermore, it can be seen from Fig. 8 that at first the processing time obtained by EA with IR is less than that by EA with IG, but as workload size increases, EA with IG outperforms EA with IR. This means that with increasing workload size, the influence of processor release times on the processing time becomes weaker, while the influence of load distribution sequence on processing time becomes stronger.

4.2 Evaluating the Performance of PAA-GA

By taking processor available time periods into account, we make a comparison between the proposed PAA-GA and EA with three commonly used load distribution sequences as its input: IG, IW, and IR. Given that the original EA proposed in [34] does not consider processor off-line times, those processors whose processing times exceed their off-line times need to be rescheduled. For simplicity, we reallocate those load fractions to the processor with the largest off-line time. Figure 9 shows the variation of processing time obtained by PAA-GA, EA-IG, EA-IW, and EA-IR along with different workload size. As shown in Fig. 9, for each workload,

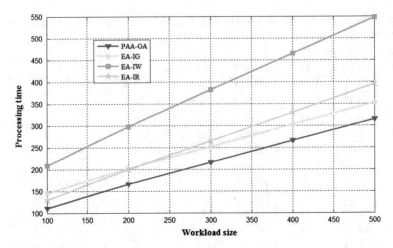

Fig. 8 Variation of processing time obtained by PAA-GA, EA-IG, EA-IW, and EA-IR along with different workload sizes for processor release-time aware divisible-load scheduling

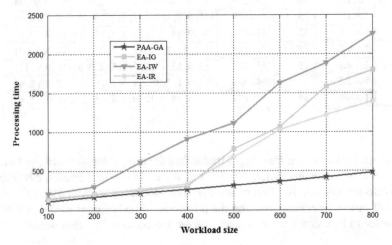

Fig. 9 Variation of processing time obtained by PAA-GA, EA-IG, EA-IW, and EA-IR along with different workload sizes for processor available-time aware divisible-load scheduling

the processing time obtained by PAA-GA is much less than that by EA with any of the load distribution sequences, and the time difference between them grows with increasing workload size. When workload size is as large as 800 in our experiment, the processing time obtained by PAA-GA shows a gain of 64% compared to EA with IR, gained about 72% compared to EA with IG, and gained over 77% compared to EA with IW. Therefore, it is clear that although the effect of available times has a greater influence on the performance as workload size increases, PAA-GA outperforms over other strategies as testified in our evaluations.

5 Conclusions

One of the key characteristics of sustainable computing systems is in efficiently managing available shared resources by designing judicious scheduling algorithms. By designing optimal, if not time-efficient scheduling algorithms, energy consumption is indirectly managed. Towards this effort, in this chapter, we have proposed an efficient processor availability-aware scheduling model to optimize the energy efficiency of heterogeneous sustainable computing systems. Using this model, we designed a genetic algorithm based global optimization strategy to derive an optimal load partition together with an optimal distribution sequence. This is an important contribution to the literature as this is the first time where the modeling is tuned to accommodate all influencing parameters (start-up overheads, processor availabilities, heterogeneous networks) to achieve a global optimal solution. We have conducted several experiments to demonstrate the correctness and effectiveness of the proposed algorithm PAA-GA. Experimental results showed that although the effect of processor available time periods has a greater influence on the performance as the workload size increases, the proposed PAA-GA reduced the processing time, hence the energy consumption too, by over 60% compared to other strategies. An important and an immediate useful extension to the study posed in this chapter is in developing a similar strategy for an arbitrary topology as real-life network based computing platforms seldom have regular topologies.

Acknowledgements This is a collaborative research work conducted jointly between Department of Electrical and Computer Engineering, National University of Singapore, Singapore, and School of Computer Science and Technology, Xidian University, China and supported by National Natural Science Foundation of China (No. 61402350, No. 6 1472297, and No. 61572391), the Fundamental Research Funds for the Central Universities (No. JB150307) and China Scholarship Council.

References

1. L. Mashayekhy, M.M. Nejad, D. Grosu, Q. Zhang, W. Shi, Energy-aware scheduling of mapreduce jobs for big data applications. IEEE Trans Parallel Distrib. **26**, 2720–2733 (2015)
2. J. Cao, K. Li, I. Stojmenovic, Optimal power allocation and load distribution for multiple heterogeneous multicore server processors across clouds and data centers. IEEE Trans. Comput. **63**, 45–58 (2014)
3. Y.N. Xia, M.C. Zhou, X. Luo et al., A stochastic approach to analysis of energy-aware DVS-enabled cloud datacenters. IEEE Trans. Syst. Man Cybern. A **45**, 73–83 (2015)
4. X. Zhu, C. He, K. Li et al., Adaptive energy-efficient scheduling for real-time tasks on DVS-enabled heterogeneous clusters. J. Parallel Distrib. Combut. **72**, 751–763 (2012)
5. G. Terzopoulos, H. Karatza, Performance evaluation and energy consumption of a real-time heterogeneous grid system using DVS and DPM. Simul. Model Pract. Theory **36**, 33–43 (2013)
6. P. Zhou, W. Zheng, An efficient biobjective heuristic for scheduling workflows on heterogeneous DVS-enabled processors. J. Appl. Math. **2014** (2014)
7. B. Luo, S. Wang, W. Shi, Y. He, eCope: workload-aware elastic customization for power efficiency of high-end servers. IEEE Trans. Cloud Comput. **4**, 237–249 (2016)

8. P. Lama, Y. Guo, C. Jiang, X. Zhou, Autonomic performance and power control for co-located web applications in virtualized datacenters. IEEE Trans. Parallel Distrib. **27**, 1289–1302 (2016)
9. Z. Xiao, W. Song, Q. Chen, Dynamic resource allocation using virtual machines for cloud computing environment. IEEE Trans. Parallel Distrib. **24**, 1107–1117 (2013)
10. U. Wajid, P. Plebani, B. Pernici et al., On achieving energy efficiency and reducing CO_2 footprint in cloud computing. IEEE Trans. Cloud Comput. **4**, 138–151 (2016)
11. H. Liu, H. Jin, C.Z. Xu et al., Performance and energy modeling for live migration of virtual machines. Cluster Comput. **16**, 249–264 (2013)
12. A. Corradi, M. Fanelli, L. Foschini, VM consolidation: a real case based on OpenStack Cloud. Future Gener. Comput. Syst. **32**, 118–127 (2014)
13. X. Dai, M. Wang, B. Benasou, On achieving energy efficiency and reducing CO_2 footprint in cloud computing. IEEE Trans. Cloud Comput. **4**, 210–221 (2016)
14. E. Feller, L. Ramakrishnan, C. Morin, Performance and energy efficiency of big data applications in cloud environments: a Hadoop case study. J. Parallel Distrib. Comput. **79**, 80–89 (2015)
15. T.G. Robertazzi, Ten reasons to use divisible load theory. Computer **36**, 63–68 (2003)
16. S. Momcilovic, A. Ilic, N. Roma, L. Sousa, Dynamic load balancing for real-time video encoding on heterogeneous CPU+GPU systems. IEEE Trans. Multimedia **16**, 108–121 (2014)
17. A. Ilic, S. Momcilovic, N. Roma, L. Sousa, Adaptive scheduling framework for real-time video encoding on heterogeneous systems. IEEE Trans. Circ. Syst. Video Technol. **26**, 597–611 (2016)
18. S. Suresh, H. Huang, H.J. Kim, Scheduling in compute cloud with multiple data banks using divisible load paradigm. IEEE Trans. Aerosp. Electron. Syst. **51**, 1288–1297 (2015)
19. Z. Ying, T.G. Robertazzi, Signature searching in a networked collection of files. IEEE Trans. Parallel Distrib. **25**, 1339–1348 (2014)
20. V. Mani, D. Ghose, Distributed computation in linear networks: closed-form solutions. IEEE Trans. Aerosp. Electron. Syst. **30**, 471–483 (1994)
21. T.E. Carroll, D. Grosu, Strategyproof mechanisms for scheduling divisible loads in bus-networked distributed systems. IEEE Trans. Parallel Distrib. **19**, 1124–1135 (2008)
22. S. Ghanbari, M. Othman, M.R.A. Bakar, W.J. Leong, Multi-objective method for divisible load scheduling in multi-level tree network. Future Gener. Comput. Syst. **54**, 132–143 (2016)
23. Z. Zhang, T.G. Robertazzi, Scheduling divisible loads in Gaussian, mesh and torus network of processors. IEEE Trans. Comput. **64**, 3249–3264 (2015)
24. C.Y. Chen, C.P. Chu, Novel methods for divisible load distribution with start-up costs on a complete b-Ary tree. IEEE Trans. Parallel Distrib. **26**, 2836–2848 (2015)
25. K. Wang, T.G. Robertazzi, Scheduling divisible loads with nonlinear communication time. IEEE Trans. Aerosp. Electron. Syst. **51**, 2479–2485 (2015)
26. V. Bharadwaj, D. Ghose, V. Mani, Optimal sequencing and arrangement in distributed single-level tree networks with communication delays. IEEE Trans. Parallel Distrib. **5**, 968–976 (1994)
27. H.J. Kim, G.I. Jee, J.G. Lee, Optimal load distribution for tree network processors. IEEE Trans. Aerosp. Electron. Syst. **32**, 607–612 (1996)
28. B. Veeravalli, X. Li, C.C. Ko, On the influence of start-up costs in scheduling divisible loads on bus networks. IEEE Trans. Parallel Distrib. **11**, 1288–1305 (2000)
29. S. Mingsheng, Optimal algorithm for scheduling large divisible workload on heterogeneous system. Appl. Math. Model. **32**, 1682–1695 (2008)
30. V. Bharadwaj, G. Barlas, Scheduling divisible loads with processor release times and finite size buffer capacity constraints in bus networks. Cluster Comput. **6**, 63–74 (2003)
31. M. Gallet, Y. Robert, F. Vivien, Comments on "design and performance evaluation of load distribution strategies for multiple loads on heterogeneous linear daisy chain networks". J. Parallel Distrib. Comput. **68**, 1021–1031 (2008)
32. M. Hu, B. Veeravalli, Requirement-aware strategies with arbitrary processor release times for scheduling multiple divisible loads. IEEE Trans. Parallel Distrib. **22**, 1697–1704 (2011)

33. S. Suresh, V. Mani, S.N. Omkar, H.J. Kim, A real coded genetic algorithm for data partitioning and scheduling in networks with arbitrary processor release time, in *Asia-Pacific Computer Systems Architecture Conference* (2005), pp. 529–539

34. K. Choi, T.G. Robertazzi, An exhaustive approach to release time aware divisible load scheduling. Int. J. Internet Distrib. Comput. Syst. **1**, 40–50 (2011)

35. H.J. Kim, A novel optimal load distribution algorithm for divisible loads. Cluster Comput. **6**, 41–46 (2003)

36. D.C. Snowdon, S.M. Petters, G. Heiser, Accurate on-line prediction of processor and memory energy usage under voltage scaling, in *7th ACM and IEEE International Conference on Embedded Software* (2007), pp. 84–93

37. K. Li, X. Tang, K. Li, Energy-efficient stochastic task scheduling on heterogeneous computing systems. IEEE Trans. Parall. Distrib. **25**, 2867–2876 (2014)

38. D.M. Bui, H.Q. Nguyen, Y. Yoon, S. Jun, M.B. Amin, S. Lee, Gaussian process for predicting CPU utilization and its application to energy efficiency. Appl. Intell. **43**, 874–891 (2015)

39. J.H. Holland, Adaptation in natural and artificial systems: an introductory analysis with applications to biology, control, and artificial intelligence. U. Michigan Press. (1975)

40. L. Hernando, A. Mendiburu, J.A. Lozano, A tunable generator of instances of permutation-based combinatorial optimization problems. IEEE T. Evolut. Comput. **20**, 165–179 (2016)

41. B.W. Goldman, W.F. Punch, Analysis of Cartesian genetic programming's evolutionary mechanisms. IEEE Trans. Evolut. Comput. **19**, 359–373 (2015)

Efficiency in Energy Decision Support Systems Using Soft Computing Techniques

Konstantinos Kokkinos, Elpiniki Papageorgiou, Vassilios Dafopoulos and Ioannis Adritsos

Abstract In the recent years, the advent of globalization has caused a steep rise in energy consumption especially for the most developed countries. At the same time, there is an important need for energy planning especially from the energy distributors' point of view as, it is vital for setting energy wholesale prices and profit margins. The same argument also holds for the case of energy transition from fossil fuels to more environmentally friendly forms since, it is necessary to predict future energy demands in order to achieve sustainability by balancing energy supply and demand. Even though a lot of research has been produced towards the development of energy models that try to optimize sustainability, we claim that, there is a need for multi-criteria Decision Support Systems, (DSS) that integrate a variety of econometric and computational intelligence methodologies in order to evaluate the impacts of a mix of sometimes conflicting factors such as climatological conditions, global prices, availability etc. In this paper, we propose an integrated DSS that involves Adaptive Neuro Fuzzy Systems (ANFIS), Neural Networks, (NN) and Fuzzy Cognitive Maps, (FCM) along with Econometric Models, (EM) in a hybrid fashion to predict energy consumption and prices. Such a system defines the basic structure in building energy networks that use combined heat and power. Even though the system can be applied for the totality of energy sources, we focus only on the prediction methodologies for the natural gas consumption for the country

K. Kokkinos (✉)
Computer Science Department, University of Thessaly, Lamia, Greece
e-mail: kokkinos@uth.gr; konst.kokkinos@gmail.com

E. Papageorgiou
Computer Engineering Department, Technological Educational
Institute of Central Greece, 35100 Lamia, Greece
e-mail: epapageorgiou@teiste.gr

V. Dafopoulos · I. Adritsos
Electrical Engineering Department, Technological Educational
Institute of Thessaly, 41110 Larissa, Greece
e-mail: dafopoulos@teilar.gr

I. Adritsos
e-mail: adritsos@teilar.gr

© Springer International Publishing AG 2017
A.K. Sangaiah et al. (eds.), *Intelligent Decision Support Systems
for Sustainable Computing*, Studies in Computational Intelligence 705,
DOI 10.1007/978-3-319-53153-3_3

level. We provide a comparative analysis of the aforementioned methods by obtaining the MSA, RMSE, MAE and MAPE errors, as well as R^2 metric.

Keywords ANFIS · Neural networks · Fuzzy cognitive maps · Energy consumption · Econometric model

1 Introduction

Energy consumption and administration plays a significant role in countries, especially for their development. In the last decades, all countries have tremendously increased the consumption of fossil fuels (mostly coal and oil) as the primary source for electricity, heating, transportation and factory energy covering. According to International Energy Agency (IEA) data from 1990 to 2015, the energy consumption per capita increased by 11% while, the world population increased by 29% [1], due to the huge population growth of India, China and Middle East.

Even though the energy consumption in the G20 countries has slowed down, there has been an overall worldwide increase which directly affects the global warming. The last decades and under this present situation, natural gas has become a challenging energy form and at the same time very popular as, it performs better in producing higher heating power than the most common fossil fuels. Most importantly, the carbon pollution produced by natural and shale gas is far less than coal and oil as it burns cleaner than the other two fuels. Even though the current technology in alternative energy sources, especially in renewable energy, is advanced, the majority of global energy consumption still relies heavily on non-renewable sources. At present, natural gas has already played a significant role in global energy consumption, especially in North America and Europe.

The natural gas production, distribution and consumption has been captured by a sophisticated supply chain process including not only manufacturers and suppliers, but also transporters, warehouses, retailers, and major customers themselves. It has affected the international market and therefore, it has become a great challenge for the researchers to investigate the flow of information, the knowledge and the product characteristics in order to achieve the optimal management of these entities and to maximize the overall profitability in the supply chain process. However one of the most important factors in increasing profitability is for the energy distributors to make the right decisions in time depending on demand information to enhance the commercial competitive advantage in this constantly fluctuating market.

Another very important issue that affects the natural gas consumption in several countries is the specific legislation enforced that concerns the transportation and the distribution of natural gas. More specifically, in such countries [2], distributors and transporters are obliged to pre-estimate in advance each day the consumption of the next day. Furthermore, this estimation cannot exceed a deviation of 2.5% so that, the amount of purchased gas, placed in the transmission network does not affect the supply to end consumers or safe operation of the national transmission system.

The main objective of this work is to propose an integrated DSS for energy efficiency that captures historical demand data along with average meteorological data on a country level and forecasts next period demand. To succeed this goal, we proposed and evaluated new composite forecasting mechanism for energy demand and especially for natural gas demandwhich is inspired by artificial intelligence approaches such as adaptive neuro-fuzzy inference system techniques, artificial neural networks and fuzzy cognitive maps, as well as an econometric model for predicting energy prices. More specifically, we set up a prototype Decision Support System (DSS) which captures historical demand data in order to forecast next day natural gas demand on a country level. The prototype includes three soft computing methodologies and one econometric model to derive the forecasted values. To evaluate the effectiveness of each methodology, we provide a comparative analysis of the aforementioned methods by obtaining the MSA, RMSE, MAE and MAPE errors on the forecasted values obtained. We have tested the cases of three different countries, namely Austria, France and Bulgaria. These countries are chosen because of their different climatological conditions, different location in Europe and thus different access preconditions to the main pipelines coming from Russia and Eastern Asia and also different GDP level. However, the methodologies chosen to predict next day natural gas demand seem able to capture the seasonal and meteorological characteristics for all three use cases.

The rest of the paper is organized as follows: In Sect. 2 we present a thorough literature review of the application of various related soft computing techniques for the forecasting of energy, electricity or natural gas demand. In Sect. 3 we illustrate the methodologies of (a) adaptive neuro fuzzy inference systems, (b) adaptive neural networks, (c) fuzzy cognitive maps and (d) econometric models for the forecasting of demand in natural gas given time series of meteorological and social inputs. We also show the architecture of the proposed DSS prototype for the evaluation of the aforementioned techniques and we elaborate on the hardware and software idiosyncrasies of the system. In Sect. 4 we present the results of the application of the above methodologies and we depict comparative tables and figures for the calculated performance metrics on the forecasted values obtained. Finally, we conclude this paper with an overall evaluation of our approach and future challenges.

2 Literature Review

There is a considerable amount of research literature that deals with the prediction of energy consumption and specifically natural gas consumption. This research can be classified into various classes according to different criteria of taxonomy. The first classification regards the timeseries' steps used for investigation due to time-series input data used having yearly and monthly predictions to be of primary preference. A second classification refers to the investigation of several locations/areas on the world having focused on the (a) world level, (b) state level, (c) city/county regional level and (d) household level. Finally, a third classification depends on the general techniques/methodologies used for modeling the prediction process (macro and micro econometric, artificial intelligence and statistical models). Soldo in [3] has published a state-of-the-art survey of forecasting natural gas consumption focusing on the provision of analysis and synthesis of the research to achieve usable results. However, because of our concentration into models of artificial intelligence and soft computing we provide a thorough overview of this subclass of related research.

In two consecutive works, Khotanzad and Elragal [4] and Khotanzad et al. [5] explore the usage of adaptive neural networks, (ANN) in predicting consumption. More specifically, [4, 5] use a combination of 3 ANN's: (a) a multilayer feedforward network trained with backpropagation, (b) another multilayer feedforward network trained with Levenberg-Marquardt algorithm, and (c) a one-layer functional link network. These separate forecasts are nonlinearly combined in the second stage using a functional link to improve prediction accuracy. Also similar work has been presented by Brown et al. in [6, 7] where the authors use ANN to predict NG demand and they justify that ANN prediction models perform with greater accuracy than the linear regression model in the investigation of daily gas consumption prediction. The research was applied in two regions of the state of Wisconsin, US but their results only refer to specific time periods of the heating season and not whole yearly timeseries. Similar work is also presented by Ozturk et al. [8] with the main objective to develop a systematic way to establish input estimation equations for the residential-commercial sector to help them predict future natural gas amounts based on genetic algorithm (GA) for the region of Turkey. Using the same philosophy and methodology, Ivezic [9] developed a multilayer ANN model for short term natural gas consumption forecasting incorporating historical weather and consumption data and through the use of a Levenberg-Marquardt training algorithm. Experimental verification of the system was done with the use of real data of specific urban locations providing efficient results based on well-defined criteria. The last work regarding ANN and natural gas consumption prediction shown is a work by Ekici, [10] which tackles the building energy need prediction problem. The authors present a multi-criteria approach taking into consideration various building characteristics (orientation, insulation thickness and transparency). The backpropagation neural network produced gives promising forecasting results.

Recently, Azadeh et al. [11] proposed a Neuro-Fuzzy-Stochasticapproach for long-term natural gas consumption forecasting and behavior analysis. This research attempts to integrate fuzzy data especially in the case when these are noisy and of uncertain nature using a system-auto regression-analysis of variance algorithm for long term predictions of natural gas for the Iranian region. Prior to this work, Azadeh et al. in [12] did a similar research for the same region but, the investigation was for short-term predictions. The important outcome was that for this case ANFIS is more accurate than ANN by adding an additional input in ANFIS to be the same day consumption in the previous year. On the other hand, Behrouznia et al. [13] used six different ANFIS models in order to find the best fitting one while [12] compared ANFIS model with ANNmodel. Similar research lately was published by Dalfard et al. [14] with a proposition of an integrated ANFIS to forecast long-term natural gas consumption when prices experience large increase. The authors use linear regression to construct a first order fuzzy inference system. However, due to small historical data sets this method experiences random uncertainty and for that reason the authors used Monte Carlo simulations to generate training data for the proposed ANFIS. The latest work in ANFIS usage for natural gas consumption prediction is published by Wang et al. [15] presenting a review of natural gas forecasting models for the Chinese region. The authors employed the multicycle Hubbert model for prediction in production to determine the peak production, the peak year and the future production trends based on several scenarios.

On the other hand, natural gas is one of the most promising types of energy for the whole Europe in the last decade and for that reason is considered as one of the most vital factors for the economic growth and development of all countries that use it. The vast majority of the established econometric models correlate the energy demand with macroeconomic variables. Such variables include the Gross National Product, the energy price, the technology, and the population of each country. Early works like in Arsenault et al. [16] incorporate the previous input variables with the previous year's energy demand, the real income, and the heating days in regions of Canada to predict the next year demand. Similar models take also into account all energy types like electricity, oil, gas, coal, total energy demand, and technological to forecast the energy demand in the UK and Germany, [17], Italy, [18] and Greece, [19]. The most recent related work appears in [20] for the case of India. The authors determine the best fit in the curves calculated by historical data using R^2, square error (SE), and Durbin Watson (DW) statistics. More specifically, the 't' test is conducted to find the significant variables influencing energy demand and they found that the coal price and the population were the most important factors that influence the energy demand but these factors are also dependent on the GNP and electricity price.

As for the case of decision support system development, lately there has been a significant progress regarding DSS-supported business models for energy country operators and distributors, but there is always space for improvements [21], especially when these models are related to smart grids and distributed energy production. However, most of the existed systems deal with energy management and not with forecasting future demands. Only lately, there are few attempts to integrate smart grid approaches of larger public sites for forecasting future loads (either macroscopic or microscopic time periods) and for direct control of large loads and generation capabilities. When this can be succeeded, the resulted synchronization between the DSS capabilities and the task support needs can improve DSS utilization, decision maker performance, and thereby task outcomes [22, 23].

3 Materials and Methods

In this section we illustrate all the participating methodologies for forecasting the demand of natural gas either by using soft computing techniques or an econometric model based on socioeconomic variables. The following subsections elaborate on the specifics of each method. The first subsection demonstrates the logical architecture of the proposed DSS prototype and the interconnection of the artificial intelligence and socioeconomic components.

All natural gas consumption data used for the studies come from the Gas Infrastructure Europe (GIE) which is an association representing the sole interest of the infrastructure industry in the natural gas business such as Transmission System Operators, Storage System Operators and LNG Terminal Operators. GIE has currently 67 members in 24 European countries and its internal structure has three columns corresponding to the three types of infrastructure activities represented: GTE (Gas Transmission Europe), GSE (Gas Storage Europe) and GLE (Gas LNG Europe), all of which fall under the umbrella of GIE. All data are free to download from [24]. On the other hand, all meteorological data used in the study came from the NOAA National Centers for Environmental Information [25] but due to the inconsistency of the morphology separate web services were developed to integrate these data to our repository.

3.1 DSS Logical Architecture

The purpose of this research is to develop an energy monitoring and management DSS to help natural gas operators and utilities at the country level to make better and more sustainable and optimized demand management decisions. This DSS will help also the utilities to provide a financial evaluation of their management decisions while considering the supply, operational, and demand side of this energy management. We mostly concentrate on the management of natural gas distribution at a country level using soft computing and socioeconomic methodologies.

There are many drivers for operators and distributors to improve their management of energy use and demand forecasting including: (a) the energy cost management and the risk of unknown future energy cost and availability/reliability, (b) the provision of customer quality of service and (c) the understanding how future systems changes affect energy use. For the aforementioned reasons energy (and specifically natural gas) utilities need a coherent strategy that can be developed by evaluating the options in terms of meeting their stated goals with the use of an integrated DSS.

Figure 1 portrays a schematic diagram showing the proposed DSS. The proposed logical architecture consists of three major parts: (a) the data aggregation and storage subsystem, (b) the model repository and (c) the user interface. The data aggregation and storage subsystem consists of the web services used to gather all

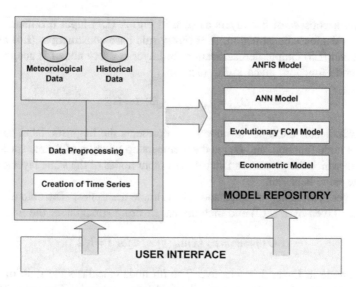

Fig. 1 Logical architecture of the proposed DSS

historical data of natural gas consumption at the country level as well as the services from national meteorological databases for the aggregation of meteorological data. The model repository is an open-ended subsystem where the developers can put all relevant components for prediction. In this work we only use artificial intelligence and socioeconomic models but the system is designed to accept statistical components or components of other nature. Finally, the user interface subsystem basically integrates the other two by a set of user friendly forms to select, insert and manage data and also a set of utilities to visualize the prediction results in a form of time series and curves.

3.2 Adaptive Neuro Fuzzy Inference Systems

AnANFIS is an inference model that maps the inter-correlation between input and output timeseries. For ANFIS this mapping uses fuzzy logic and thus is called Fuzzy Inference [26]. These systems have proved to work better when the input and output sets are time series data of a predefined time step. The rule based inference system established in an ANFIS consists of: (a) a rule-base, containing fuzzy if–then rules, (b) a data-base, defining the membership functions (MF) and (c) an inference system, combining the fuzzy rules and producing the system results as [27] justifies. There are two types of popular FIS, the Takagi–Sugeno, [28] and the Mamdani ANFIS, [29]. The two systems differ only on the definition of the consequent parameters in the network. Most popular approach is the Takagi and Sugeno type according to which, the rule base is constructed from the input–output pairs and it consists of five layers as seen in Fig. 2: (L1) Input fuzzification, (L2) Fuzzy set database construction, (L3) Fuzzy rule base construction, (L4) Decision making and (L5) Output defuzzification. In Layer 1, every node i is adaptive node with a node function, given in Eq. (1):

$$O_{1,i} = \mu_{A_i}(x) \quad for\ i = 1, 2 \tag{1}$$

where x indicates the input to node i, A_i represents the linguistic label associated with this node function, and $O_{1,i}$ is the membership function of A_i that specifies the degree to which the given x satisfies A_i. All other nodes of the same layer are set to have the same behavior.

In Layer 2, all nodes are fixed and act as simple multipliers. The outputs of these nodes are given by Eq. (2) and they are called firing strengths of the rules:

$$O_{2,i} = w_i = \mu_{A_i}(x)\mu_{B_i}(y) \quad for\ i = 1, 2 \tag{2}$$

Each node in Layer 3 is adaptive. The ith node calculates the ratio of the i-th rule's firing strength to the sum of all rules' firing strengths. Equation (3) shows how to obtain the output of this layer:

$$O_{3,i} = \overline{w}_i = \frac{w_i}{w_1 + w_2} \tag{3}$$

Each node, in Layer 4, is also an adaptive node with a function given by Eq. (4):

$$O_{4,i} = \overline{w}_i f_i = \overline{w}_i (p_i x + q_i y + r_i) \tag{4}$$

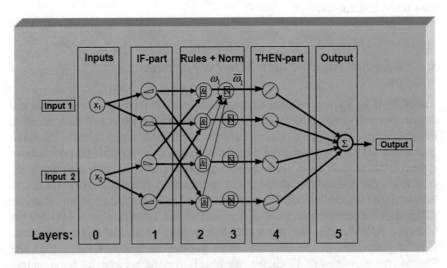

Fig. 2 Adaptive neuro-fuzzy inference system structure [32]

where $\overline{w}_i$ is the output of layer 3, and $\{p_i,\ q_i,\ r_i\}$ are referred to as consequent parameters. Finally, one node, in Layer 5, is a fixed node to be the sum of all incoming inputs:

$$O_{5,i} = \sum_i \overline{w}_i f_i = \frac{\sum_i w_i f_i}{\sum_i w_i} \tag{5}$$

In order for the ANFIS to work, we first adjust the membership function parameters using either a back propagation algorithm alone or in combination with a least squares type of method [30]. The most popular membership functions used in the ANFIS application are shown in Fig. 3.

Learning is similar to that of neural networks thus,ANFIS uses the benefits of ANN and fuzzy inference [31]. Propagation and hybrid are two learning methods which are generally applied in ANFIS to clearly describe the relationship between input and output [32]. Hybrid learning, which is a combination of gradient decent method and least squares approach, can decrease the complexity of the algorithm and simultaneously increase the learning efficiency. The parameters associated with the membership functions will change through the learning process using a gradient vector that facilitates in this recalculation. So every time the gradient vector is obtained, an optimization procedure can be performed to adjust parameters in order to reduce errors.

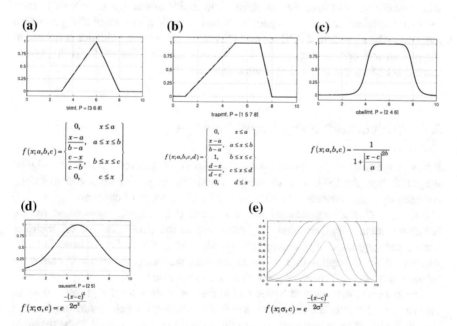

Fig. 3 Membership functions used in the ANFIS: **a** trimf, **b** trapmf, **c** gbelmf, **d** gaussmf and **e** gauss2mf

3.3   Artificial Neural Networks (ANN)

An ANN is a system of interconnected units known as neurons interacting across weighted connections. Inspired by the architecture of the human brain, these neurons can compute output values from inputs by learning complex patterns of information and generalizing the learned information [33].

ANNs can be classified into several categories based on supervised and unsupervised learning methods and feedforward and feedback recall architectures. Multilayered feedforward networks use a supervised learning method and feedforward architecture. A backpropagation neural network is one of the most frequently utilized neural network techniques for classification and prediction [34].

To be successful an ANN in prediction and classification tasks, proper selection of ANN modeling parameters is needed. The most important parameters need to be defined for ANN performance are: the number of nodes (neurons) and layers, the nonlinear function used in the nodes, the learning algorithm, the initial weights of the inputs and layers, and the number of epochs for which the model is iterated [35, 36]. In general, the ANN model has a structure of an input layer, a hidden layer, and an output layer. The input data are divided into learning and testing sets as described in [33]. The neural network uses initially the learning sets to learn the relationship between the output and input criteria, while the test set is used to assess the performance of the model during the testing process.

ANNs have been suggested as an alternative to time series forecasting to deal with linear and nonlinear relationships. The major advantage of ANNs is their flexible nonlinear modeling capability where the data are fitted for prediction purposes. The approach of ANN is data-driven, no theoretical guidance is needed to suggest an appropriate data generating process therefore the model is adaptively formed based on the features presented from the data.

3.4   Evolutionary Fuzzy Cognitive Maps

Fuzzy cognitive maps (FCM) are graph-based structures, describing signed weighted digraphs [37], able to handle vagueness or partial truth and modeling complex dynamic systems. The structure of FCMs consists of concepts (i.e., nodes) C_1, C_2, ..., C_i and connections between them (i.e., edges), represented by the adjacency matrix W. The edges W_{ij} displayed in the dimensions of the matrices denote the degrees of the causal relationship and typically lie between $[-1, 1]$, whereas $W_{ij} > 0$ implies that C_i increases with increasing C_j, $W_{ij} = 0$ means no relation and, $W_{ij} < 0$ means C_i decreases with increasing C_j.

The concepts are mapped to the real-valued activation level C_i takes values in interval $[0, 1]$ that is the degree to which the observation belongs to the concept (i.e., the value of the fuzzy membership function). As a consequence of the dynamic influence of connected nodes, the concept's state changes over time. The reasoning is performed as the calculation of Eq. (6):

$$C_i(t+1) = f\left(C_i(t) + \sum_{\substack{j=1, \\ i \neq j}}^{n} C_j(t) \cdot W_{ji} \right) \qquad (6)$$

where $f()$ stands for a sigmoid function (7), ensuring that the concept's defined value falls within interval [0, 1] and $f()$ is given by:

$$f(x) = \frac{1}{1 + e^{-cx}} \qquad (7)$$

where c—parameter, $c > 0$,

FCMs have a significant ability of learning their structure by historical data. For the FCM learning process, population-based algorithms are mostly used due to their performance on defining the structure of models [38–40]. The objective of the evolutionary learning is to optimize the matrix W with respect to the forecasting accuracy. The populations of candidate FCMs are iteratively evaluated with the use of fitness function given in Eq. (8):

$$fitness_p(J_l) = \frac{1}{a \cdot J(l) + 1} \qquad (8)$$

where a—parameter, $a > 0$, p—the number of chromosome, $p = 1, \ldots, P$, P—the population size, l—the number of population, $l = 1, \ldots, L$, L—the maximum number of populations, $J(l)$—the learning error function.

One of the most well-known population-based learning algorithm for FCM is real coded genetic algorithm RCGA [38, 40]. RCGA defines the learning error function as follows (9):

$$J(l) = \frac{1}{n(T-1)} \sum_{t=1}^{T-1} \sum_{i=1}^{n} (Z_i(t) - C_i(t))^2 \qquad (9)$$

where: t—discrete time of learning, T—the number of the learning records, n—the number of the output concepts, $C_i(t)$ is the value of the i-th output concept, $Z_i(t)$ is the reference value of the i-th output concept.

Each population of chromosomes evolves with time by operator selection, crossover and mutation. In the analysis a roulette wheel method of selection, one-point crossover and random mutation were used.

In what follows, the steps of the Real Coded Genetic Algorithm (RCGA) are given [41, 42]:

1. Initialization of the FCM model (structure of the model)
2. Input of the normalized historical learning data for a given period of time
3. Definition of the fitness function, selection strategy, crossover and mutation operators
4. Initialization of the population
5. Applying selected genetic algorithm to generate new population on the basis of selected selection strategy and genetic operators
6. Calculation of fitness function
7. If stop criterion is met go to step 8; else go to step 5
8. Input of the normalized testing or current data
9. Calculation of forecasting values and forecasting accuracy measures

3.4.1 Econometric Model for Natural Gas Forecasting

Natural gas modeling turns to be an important characteristic in the international industry for planning energy storage and optimizing costs (minimization). Furthermore, this industry is vulnerable to various political and economic decisions that sometimes lead to crises and therefore, forecasting and planning facilitates the establishment of policies and the strategic decision making. For that reason, almost all European countries need to run natural gas demand analytics to optimize the overall fluctuation of prices, balance the interdependence between oil, coal and gas and finally minimize the environmental footprint which is affected from the other two choices (coal, oil) in a burgeoning economy.

The most popular econometric models for demand or consumption forecasting first try to discover the interrelation between the participating socioeconomic variables and then to establish a mathematical formula that calculates demand as a function of the input variables. This formula is produced by introducing the Best Fit equation for demand obtained using time series models. For this present study we use the methodology of [20]. This model is a yearly forecasting methodology since most of the input variables have socioeconomic impact which cannot be identified at a daily level (e.g., Gross national product, population etc.). This macroscopic evaluation of the demand mostly fits with the policy making at a level of industry of country government and it involves the following variables (Table 1).

Table 1 Variables for econometric model

D_{best}	Best fit equation for demand obtained using time series models produced by combining the socioeconomic variables
D_{t-1}	The overall demand/consumption for the previous year
P_i	Wholesale price index (i = coal, oil and gas, electricity)
Pop	Population
GNP	Gross National Product Index
Y	Forecasted natural gas demand

The authors in [20] use a combination between additive and multiplicative models (totally 92 models) from where they pick the top ten. We take the same models used in order to compute comparative results and according to the ranking for R^2, SE, and Durbin Watson (DW) statistic. Among the top 10 models selected, "t" test was performed.

4 Results and Discussion

The three soft computing approaches proposed for forecasting were implemented in the provided datasets regarding natural gas withdrawal prediction for each one of the three countries, Austria, Bulgaria and France. The input variables of the forecasting models were five: mean temperature, wind, day of previous year, one-day before and two-days before. The predicted variable was withdrawal of natural gas. The dataset of each one country consists of three years data, 2014–2016 (till July 2016). During the performance of the models, the first two years were used for learning and the last year for testing.

At first, the neuro-fuzzy algorithm of ANFIS was used for the three datasets of the provided three countries, Austria, Bulgaria and France for withdrawal prediction of natural gas. We used the Matlab 2014b-ANFIS tool named ANFIS-Editor. The tool is designed to utilize different variables and various data classification methods to achieve the minimum error between predicted values and real data. Different ANFIS configuration sets were initially investigated. These settings included the number and type of membership functions, (MF), the type of output MF, the optimization method (hybrid or back propagation) and the number of epochs as the most important adjustments in ANFIS to reach the most effective model with minimum errors. Our primary goal was to find the effect of these adjustments and their subdivisions in different combinations in order to develop these ANFIS models and compare the results. During experimentation, a large number of possible combinations of numbers of membership functions for each predictor, from 2 to 4 and types of membership functions [triangular (trimf), generalized bell-shaped (gbelmf), Gaussian (gaussmf), Gaussian combination (gauss2mf), trapezoidal (trapmf), Π-shaped (pimf) and sigmoidal (dsigmf)] were implemented [31]. After a large number of experiments, the best configuration of ANFIS was found.

The structure that gave the best fitting would fuzzificate variables x_1–x_5 into 2,2,2,2 and 2 membership functions respectively and the fuzzification would be implemented with the use of a triangular shaped function. Constant output, hybrid optimization and 100 epochs were assigned for all cases. The partitioning was implemented through the grid partitioning method.

To define forecasting accuracy of ANN models properly selected parameters were investigated constructing different ANN models which were implemented in nntool at Matlab R2014b, considering the Levenberg-Marquardt (LM) algorithm and gradient descent (GD). The LM backpropagation algorithm was shown as the most efficient one for training ANN models for prediction.

We accomplished a large number of experiments with ANN multilayered feed-forward architecture, considering different number of hidden layers, different learning rate and momentum parameters, two most efficient learning algorithms of backpropagation ANN technique (LM and GD) and number of epochs. Through the experiments, the best results of back propagation ANNs have been received for the architecture of one hidden layer, with 10 neurons, learning rate = 0.01 and momentum = 0.1, random values of initial weights, and the LM back-propagation algorithm as the learning algorithm. The optimization algorithm was selected as a conjugate gradient algorithm. The hyperbolic tangent transfer function was used in the hidden layer, and a linear transfer function was used in the output layer. The number of epochs for best configuration was 100.

Next, considering the Evolutionary-based Fuzzy Cognitive Map for daily withdrawal forecasting, the algorithm of real-coded genetic learning for FCM development from historical data was used [38, 40]. The steps of the real-coded genetic algorithm for FCM construction are described analytically in [40] and has already been used for time series prediction problems in other domains [41–44].

The parameters of the genetic algorithm were assigned on the basis of literature and numerous trials [38, 40, 41, 43]. In this case study, the values of genetic algorithm parameters were defined: size of the initial population = 100, maximum number of populations = 200, probability of crossover = 0.5, probability of mutation = 0.05, error = 0.001. Ranking selection, uniform crossover and Mühlenbein's mutation were used. To ensure the survival of the best individual in each population, elite strategy was applied.

The relations between concepts (input and output variables) were initialized with small random values. The FCM was initialized, learned and tested on the basis of the provided real withdrawal values from the three different countries. Next, using the testing set and following a similar to neural network process for one-step ahead prediction, 208 daily predicted values of withdrawal were calculated.

Daily withdrawal forecast of natural gas seems to be achievable at satisfactory accuracy as tested for numerous ANN and ANFIS structures, as well as for RCGA-FCM. To evaluate the effectiveness of these methods and select the best fitting architecture of each approach, five popular metrics have been assessed. The metrics of accuracy used for output concept (sixth variable) are the mean square error (MSE), the root-mean-square error (RMSE), the mean absolute error (MAE), the mean absolute percentage error (MAPE), used to assess the predictive power of models (9–12). We determine the best fit in the curves calculated by historical data using R^2.

$$\text{MSE} = \frac{1}{T-1} \sum_{t=1}^{T-1} (Z_o(t) - X_o(t))^2 \tag{10}$$

$$\text{RMSE} = \sqrt{\frac{1}{T-1} \sum_{t=1}^{T-1} (Z_o(t) - X_o(t))^2} \tag{11}$$

$$\text{MAE} = \frac{1}{T-1} \sum_{t=1}^{T-1} \frac{|Z_o(t) - X_o(t)|}{Z_o(t)} \qquad (12)$$

$$\text{MAPE} = \frac{100\%}{T-1} \sum_{t=1}^{T-1} \frac{|Z_o(t) - X_o(t)|}{Z_o(t)} \qquad (13)$$

where Z_o is the actual value and X_o is the forecast value.

All the results concerning the three countries for each one of the three different approaches are gathered in Tables 2 and 3. Table 3 depicts the R^2 values of the testing period for the three countries, which were calculated up to 0.9162, 0.9962 and 0.9505 respectively; this is an acceptable result for this type of prediction.

Table 2 Scatterplots and R^2 of the three approaches ANN, ANFIS and RCGA-FCM

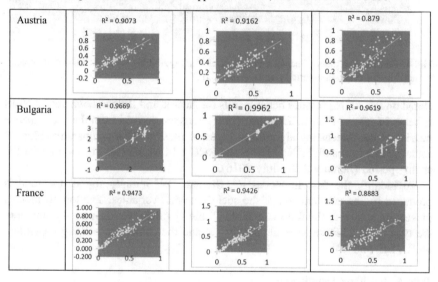

Table 3 Metrics of accuracy for the three methods

Country	Method	MSE	RMSE	MAE	MAPE	R^2
Austria	ANN	0.0046	0.0047	0.051	13.794	**0.9162**
	ANFIS	0.0046	0.0047	0.039	3.023	**0.9073**
	RCGA-FCM	0.0079	0.0062	0.0063	16.212	**0.8797**
Bulgaria	ANN	0.0011	0.0023	0.028	12.045	**0.9962**
	ANFIS	0.0060	0.0054	0.035	3.151	**0.9579**
	RCGA-FCM	0.0055	0.0051	0.048	10.43	**0.9619**
France	ANN	0.0034	0.0041	0.0416	2.196	**0.9426**
	ANFIS	0.0030	0.0038	0.037	1.1748	**0.9473**
	RCGA-FCM	0.0076	0.0061	0.064	3.349	**0.8883**

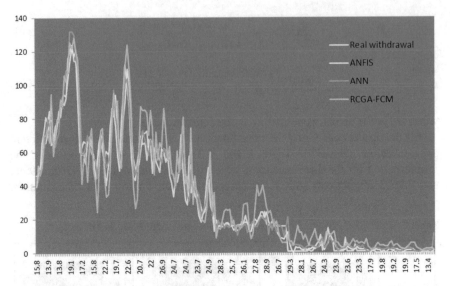

Fig. 4 Comparative plots of actual withdrawal of France, ANFIS, ANN, RCGA-FCM forecast versus time for the testing period (January–July 2016)

Based on R^2, ANN provides better results for Austria and Bulgaria, whereas ANFIS is better in case of France. The same conclusions are derived from the performance metrics of the forecasting algorithms. Figure 4 illustrates the comparative plots of actual withdrawal and ANN, ANFIS and RCGA-FCM forecast versus time for the testing period (January 2016–July 2016).

To identify the best fit model R^2, SE and DW initially for all the models produced by the combination of the socioeconomic variables. The top 10 models picked are shown in the following table (Table 4) where we illustrate the function, the type (A = additive M = multiplicative) and the composite ranking dependent on R^2, SE, and DW.

Table 4 Top 10 econometric models

Model	Type*	R^2	SE	DW
$Y = f(D_{t-1}, P_i, GNP, Pop)$	A	0.843	55176.03	2.341
$Y = f(D_{t-1}, P_i, Pop)$	A	0.845	55781.18	2.488
$Y = f(D_{t-1}, P_i, GNP)$	A	0.845	57389.65	2.113
$Y = f(D_{t-1}, P_i, GNP/Pop)$	A	0.810	57412.94	2.678
$Y = f(D_{t-1}, P_i/Pop, GNP)$	A	0.806	59680.54	1.924
$Y = f(D_{t-1}, P_i, GNP, Pop)$	M	0.837	53452.51	2.763
$Y = f(D_{t-1}, P_i, Pop)$	M	0.837	54471.80	2.646
$Y = f(D_{t-1}, P_i, GNP)$	M	0.819	54490.23	2.436
$Y = f(D_{t-1}, P_i, GNP/Pop)$	M	0.816	56597.47	1.912
$Y = f(D_{t-1}, P_i/Pop, GNP)$	M	0.816	57130.07	1.873
* A = additive M = multiplicative				

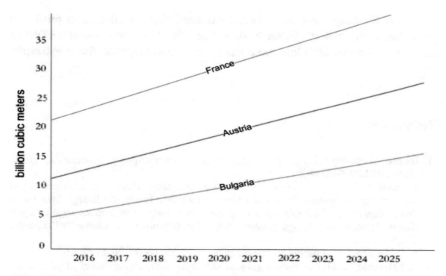

Fig. 5 Comparison between the three country predictions

We did not perform comparisons between natural gas and other sources of energy therefore the P_i proved to be insignificant variable for the study after the t-test was performed. The rest of the variables are competitive with the demand of the previous year to be the most significant in most of the cases.

The best fit economic model determined by the process above was the one in Eq. (14):

$$Y = -3421 + 0.654D_{t-1} + 21423.68(GNP/Pop) \qquad (14)$$

Finally in Fig. 5, we show the projected demands for the three countries from the year 2016 to 2025 (i.e., for the next decade). As it was all expected, the increase of the population affects positively the increase of the GNP thus the two variables are mutually interdependent and affect positively the increase on the forecasting. The same holds for the real consumption of natural gas of the previous year.

5 Conclusions

In this study, the problem of energy demand forecastingand decision making was investigated. The generic logic architecture of the DSS is defined, figuring out the main functionalities of the DSS on forecasting and the model repository as an open-ended subsystem including all relevant components for prediction. In this stage of the work, we mainly use soft computing techniques and socioeconomic models but the system is designed to accept statistical components or components of other nature. The conducted experiments showed that the three soft computing algorithms for daily withdrawal natural gas prediction are efficient in prediction performing with a high R^2. Future work is directed to the construction of the DSS of energy management including more functionality and preparing the user-friendly interface.

References

1. Energy Technology Perspectives (International Energy Agency Publications, 2016). ISBN PRINT 978-92-64-25234-9
2. SorinNeascu, Silvian Suditu and CătălinPopescu, An optimized way for a better accuracy of gas consumption profiles, Recent Researches in Electric Power and Energy Systems, in *Proceedings of the 13th International Conference on Electric Power Systems, High Voltages, Electric Machines and Energy Systems (POES '13)*, (Chania, Crete, Greece, 2013) August 27–29
3. B. Soldo, Forecasting natural gas consumption. Appl. Energy **92**, 26–37 (2012)
4. A. Khotanzad, H. Elragal, Natural gas load forecasting with combination of adaptive neural networks, in *Proceedings of the International Joint Conference on Neural Networks*, 1999, pp. 4069–4072
5. A. Khotanzad, H. Elragal, T.L. Lu, Combination of artificial neural-network fore-casters for prediction of natural gas consumption. IEEE Trans. Neural Netw. **11**(2), 464–4730 (2000)
6. R.H. Brown, P. Kharouf, F. Xin, L.P. Piessens, D. Nestor, Development of feed-forward network models to predict gas consumption, Neural Networks, in *IEEE World Congress on Computational Intelligence*, 1994, pp. 802–805
7. R.H. Brown, I. Matin, Development of artificial neural network models to predict daily gas consumption, in *Industrial Electronics, Control, and Instrumentation, Proceedings of the 1995 IEEE IECON*, pp. 1389–1394
8. H.K. Ozturk, O.E. Canyurt, A. Hepbasli, Z. Utlu, Residential-commercial energy input estimation based on genetic algorithm (GA) approaches: an application of Turkey. Energy Build. **36**(2), 175–183 (2004)
9. D. Ivezic, Short-term natural gas consumption forecast. FME Trans. **34**, 165–169 (2006)
10. B.B. Ekici, U.T. Aksoy, Prediction of building energy consumption by using artificial neural networks. Adv. Eng. Softw. **40**, 356–362 2009
11. A. Azadeh, S.M. Asadzadeh, M. Saberi, V. Nadimi, A. Tajvidi, A neuro-fuzzy stochastic frontier analysis approach for long-term natural gas consumption forecasting and behavior analysis: the cases of Bahrain, Saudi Arabia, Syria, and UAE. Appl. Energy **40**, 1432–1445 (2011)
12. A. Azadeh, S.M. Asadzadeh, A. Ghanbari, An adaptive network-based fuzzy inference system for short-term natural gas demand estimation: uncertain and complex environments. Energy Policy **38**(3), 1529–1536 (2010)
13. A. Behrouznia, M. Saberi, A. Azadeh, S.M. Asadzadeh, P. Pazhoheshfar, An adaptive network based fuzzy inference system-fuzzy data envelopment analysis for gas consumption forecasting and analysis: the case of South America, in *International conference on intelligent and advanced systems*, (2010) Article number 5716160
14. V. MajaziDalfard, M. NazariAsli, S.M. Asadzadeh, S.M. Sajjadi, A. Nazari-Shirkouhi, A mathematical modeling for incorporating energy price hikes into total natural gas consumption forecasting. Appl. Math. Model. **37**(8), 5664–5679 (2013)
15. J. Wang, H. Jiang, Q. Zhou, J. Wu, S. Qin, China's natural gas production and consumption analysis based on the multicycle Hubbert model and rolling Grey model. Renew. Sustain. Energy Rev. **53**, 1149–1167 (2016)

16. E. Arsenault, J.T. Bernard, C.W. Carr, E. Genest-Laplante, A total energy demand model of Québec. Forecast. Prop. Energy Econ. **17**, 163–171 (1995)
17. I.D. McAvinchey, A. Yannopoulos, Stationarity, structural change and specification in a demand system: the case of energy. Energy Econ. **25**, 65–92 (2003)
18. F. Gori, C. Takanen, Forecast of energy consumption of industry and household and services in Italy. Heat Technol. **22**, 115–121 (2004)
19. C.H. Skiadas, L.L. Papayannakis, A.G. Mourelatos, An attempt to improve the forecasting ability of growth functions: The Greek electric system. Technol. Forecast. Soc. Chang. **44**, 391–404 (1993)
20. L. Suganthi, Anand A. Samuel, Modelling and forecasting energy consumption in INDIA: Influence of socioeconomic variables. Energy Sour. Part B: Econ. Plan. Policy **11**(5), 404–411 (2016). doi:10.1080/15567249.2011.631087
21. SEMANCO Project Website, http://www.semanco-project.eu/
22. D.L. Goodhue, R.L. Thompson, Task-technology fit and individual performance. MIS Q. **19**(1), 213–236 (1995)
23. C. Smith, J. Mentzer, Forecasting task-technology fit: the influence of individuals, systems and procedures on forecast performance. Int. J. Forecast. **26**(1), 144–161 (2010)
24. Gas Infrastructure Europe, http://www.gie.eu/index.php/publications
25. NOAA National Centers for Environmental Information, https://www.ncdc.noaa.gov
26. V. Adriaenssens, B. De Baets, P.L.M. Goethals, N. De Pauw, Fuzzy rule-based models for decision support in ecosystem management. Sci. Total Environ. **319**, 1–12 (2004)
27. Z. Sen, Fuzzy Logic and Foundation. (Bilge KulturSanat Publisher, Istanbul, 2001), p. 172. ISBN 9758509233
28. T. Takagi, M. Sugeno, Fuzzy identification of systems and its application to modeling and control. IEEE Trans. Syst. Man Cybern. **15**, 116–132 (1985)
29. J.S.R. Jang, C.T. Sun, E. Mizutani, *Neuro-Fuzzy and Soft Computing* (Prentice Hall, 1997), p. 607 ISBN 0-13-261066-3
30. R. Singh, A. Kainthola, T.N. Singh, Estimation of elastic constant of rocks using an ANFIS approach. Appl. Soft Comput. **12**, 40–45 (2012)
31. A. Azadeh, M. Saberi, M. Anvari, A. Azaron, M. Mohammadi, An adaptive network based fuzzy inference system–genetic algorithm clustering ensemble algorithm for performance assessment and improvement of conventional power plants. Expert Syst. Appl. **38**, 2224–2234 (2011)
32. B. Khoshnevisan, S. Rafiee, H. Mousazadeh, Application of multi-layer adaptive neuro-fuzzy inference system for estimation of greenhouse strawberry yield. Measurement **47**, 903–910 (2014)
33. S. Haykin, *Neural Networks: A Comprehensive Foundation* (Prentice Hall, New Jersey, 1998)
34. G.P. Zhang, Neural networks for classification: a survey. IEEE Trans. Syst. Man and Cybern. Part C: Appl. Rev. **30**(4), 451–462 (2000)
35. C.M. Bishop, *Neural Networks for Pattern Recognition* (Oxford University Press, Oxford H. R., 1995)
36. M. Firat, M. Yurdusev, M. Turan, Evaluation of artificial neural network techniques for municipal water consumption modeling. Water Resour. Manage. **23**, 617–632 (2009)
37. B. Kosko, Fuzzy cognitive maps. Int. J. Man-Mach. Stud. **24**(1), 65–75 (1986). doi:10.1016/S0020-7373(86)80040-2
38. W. Stach, L. Kurgan, W. Pedrycz, M. Reformat, Genetic learning of fuzzy cognitive maps. Fuzzy Sets Syst. **153**(3), 371–401 (2005). doi:10.1016/j.fss.2005.01.009
39. E.I. Papageorgiou, *Fuzzy Cognitive Maps for Applied Sciences and Engineering From Fundamentals to Extensions and Learning Algorithms Intelligent Systems Reference Library*, vol. 54 (Springer, 2014)
40. E.I. Papageorgiou, W. Froelich, Multi-step prediction of pulmonary infection with the use of evolutionary fuzzy cognitive maps. Neurocomputing **92**, 28–35 (2012). doi:10.1016/j.neucom.2011.08.034

41. E.I. Papageorgiou, K. Poczeta, A. Yastrebov, Fuzzy Cognitive Maps and Multi-Step Gradient Methods for Prediction: Applications to Electricity Consumption and Stock Exchange Returns, in *7th International KES Conference on INTELLIGENT DECISION TECHNOL-OGIES KES-IDT-15*, Hilton Sorrento Palace, Italy, 17–19 June 2015, published by Springer in Smart Innovation Systems and Technologies, http://idt—15.kesinternational.org/cms/userfiles/is02.pdf

42. E.I. Papageorgiou, K. Poczeta, Application of Fuzzy Cognitive Maps to Electricity Consumption Prediction, in *NAFIPS 2015, Annual Conference of the North American Fuzzy Information Processing Society NAFIPS'2015*, Redmond, Washington, USA, August 17–19, 2015

43. W. Froelich, E.I. Papageorgiou, M. Samarinas, K. Skriapas, Application of evolutionary fuzzy cognitive maps to the long-term prediction of prostate cancer. Appl. Soft Comput. **12**, 3810–3817 (2012). doi:10.1016/j.asoc.2012.02.005

44. K. Kokkinos, E. Papageorgiou, K. Poczeta, L. Papadopoulos, C. Laspidou, Soft Computing Approaches for Urban Water Demand Forecasting, in *8th International Conference KES-IDT 2016*, Tenerife, Spain, 15–17 June, 2016, published by Springer in Smart Innovation Systems and Technologies

Computational Intelligence Based Heuristic Approach for Maximizing Energy Efficiency in Internet of Things

Amandeep Verma, Sakshi Kaushal and Arun Kumar Sangaiah

Abstract Internet of Things (IoT) is a network of physical things or objects fixed with electronics, software, sensors and network connectivity. It is a new revolution of the Internet that is hastily gathering impetus driven by the advancements in sensor networks, mobile devices, networking, and wireless and cloud technologies. It permits these things or objects to collect and transact and analyze data for performing specific tasks in digital world with limited storage and processing capacities. Due to massive adoption of cloud computing having virtually unlimited storage and processing capabilities, it can be merged with IoT to be an important component of Future Internet. This chapter explores IoT and cloud computing as well as their symbiosis based on the common environment of distributed processing. For IoT devices to be operational for longer time, the development of energy efficient schemes for sustainable computing environment is a challenging issue. It is possible to lease on-demand computing resources through cloud in an optimized manner from energy point of view. Further, Computational Intelligence can help to save device resources and energy by shifting the computational tasks from device to cloud. To make the devices energy efficient, this chapter also presents a dominance sort based optimized heuristic to offload and process local computations on cloud. The proposed approach uses the multi-objective swarm intelligence technique, i.e., Multi-Objective Particle Swarm Optimization (MOPSO) to generate Pareto optimal solutions for task offloading. Our method first determined which of the tasks can be run locally over the mobile cores and which are to be offloaded to the cloud. This will result into lower cost and high value to end user of services. The overall outcome can be helpful to bolster the performance of IoT devices. Higher operational efficiency in

A. Verma (✉) · S. Kaushal
UIET, Panjab University, Chandigarh, India
e-mail: amandeepverma@pu.ac.in

S. Kaushal
e-mail: sakshi@pu.ac.in

A.K. Sangaiah
School of Computing Science and Engineering, VIT University,
Vellore, Tamil Nadu, India
e-mail: arunkumarsangaiah@gmail.com

© Springer International Publishing AG 2017
A.K. Sangaiah et al. (eds.), *Intelligent Decision Support Systems
for Sustainable Computing*, Studies in Computational Intelligence 705,
DOI 10.1007/978-3-319-53153-3_4

system will help and support to make IoT sustainable in long run. These improvements will pave the way for achieving a new level in IoT industry and establishing new standard for benchmarking.

Keywords Computational intelligence · Sustainable computing · IoT · Cloud · Non-dominance · Pareto optimal solution

1 Introduction

Cloud computing is the latest emerging paradigm of distributed computing which uses the concept of hardware and software virtualization to provide a dynamically scalable services. Based upon the demand, these services can be accessed over Internet. In the past few years, Distributed and Parallel Computing as well as Service Oriented Computing attract the interests of researchers [1]. As compared with the other traditional computing paradigms like cluster, Grid, and peer-to-peer (p2p), the Cloud computing adopts a market-oriented business model where users are charged for consuming Cloud services such as computing, storage, and network services like conventional utilities in everyday life (e.g. water, electricity, gas, and telephony) [2]. Cloud computing delivers three defined models as *Software as a Service* (*SaaS*), where the user uses the various applications but has no control over the hosting environment. The examples of SaaS include Google Apps and Salesforce.com [3]. *Platform as a Service* (*PaaS*) offers a full or partial application development environment that users can access and utilized online collectively or individually. It facilitates the deployment of applications without the cost and complexity of buying and managing the underlying hardware and software and provisioning hosting capabilities. In this model, the platform is typically an application framework. AWS and Google App Engine are PaaS cloud providers [4]. In infrastructure as a service (*IaaS*), the service provider provides the wide variety of resources of different processing power and storage capabilities. The user can use these resources to deploy its own applications. The user no longer needs to maintain the hardware. Amazon EC2, Globus, Nimbus, and Eucalyptus are *IaaS* providers [5]. Scalability and flexibility are two main advantages of cloud where the user can access and release the resources as per their need.

On the other hand, Internet of Things (IoT) is a network of physical things or objects fixed with electronics, software, sensors and network connectivity. It is based upon ubiquitous and pervasive computing [6]. IoT consists of real world small things with limited processing and storage capacities. But, the cloud virtually provides unlimited storage and processing capabilities. Thus, by integrating these two complementary techniques, the mutual advantages have been identified in the literature and this new paradigm is known as CloudIoT [7]. In general, to handle the issues of limited processing and storage capabilities, IoT can use unlimited resources of Cloud. Similarly, Cloud can extend its scope from virtual to real world things with the help of IoT. Cloud acts as an intermediate between the real things and virtual applications. It hides the complex functionality details to implement the IoT [8]. The various applications of IoT includes Healthcare services [9], Smart cities and communities [10], Smart home [11], Video surveillance [12], and Energy efficient smart grid [13] etc. All of these applications need massive storage and computational resources. Combining IOT with Cloud can solve the problem of processing and storage. In the following we list the few advantages obtained using CloudIoT paradigm:

(i) Storage: IoT applications generate a large volume of structured and semi-structured data. It requires collecting, processing, sharing and searching of large volume of data. This problem can be solved by accessing the unlimited, cost efficient and on-demand storage services of Cloud [14].

(ii) Computational Resources: IoT devices cannot perform complex on-site data processing due to their limited processing and limited battery. So, major processing unit of an application is transmitted to nodes that are more powerful in terms of processing and storage. As, cloud provides virtually unlimited processing capabilities, this represents another important CloudIoT driver. The major processing part of an application is offloading to the cloud for energy saving of IoT devices [7].

Thus, these several motivations lead to the integration of Cloud and IoT. But, at the same time, it imposes several challenges for each application. Main challenges include heterogeneity of resources, security, performance, reliability and power and energy efficiency, etc. [8].

For IoT devices to be operational for longer time, the development of energy efficient schemes for sustainable computing environment is a challenging issue. Energy efficient task offloading is currently getting interest of the research community [15]. Computational Intelligence techniques can help to save device resources and energy by shifting the computational tasks from device to cloud. Computational Intelligence techniques generate number of Pareto optimal solutions depending upon the user requirements.

This chapter first explores the work done in the area of energy efficient task offloading techniques for IoT enabled mobile devices. From the study of work done, we have found that none of the existing techniques used Computational Intelligence techniques to give optimal energy saving results to the user. So, to make the devices energy efficient, in this chapter, we have also proposed a novel technique to offload the task from IoT enabled mobile devices to cloud. The proposed approach uses the multi-objective computational intelligence to generate Pareto optimal solutions for task offloading. Our method first determined which of the tasks can be run locally over the mobile cores and which are to be offloaded to the cloud. Then, results of this assignment is fed into the initial population of multi-objective swarm intelligence technique, i.e., Multi-Objective Particle Swarm Optimization (MOPSO) to schedule the task either over mobile cores or offloaded to cloud such that the precedence requirements among the different tasks along with time constraints are met and energy consumption of IOT mobile devices is minimized. The simulation analysis validates that the solutions obtained with proposed heuristic deliver better convergence and uniform spacing among the solutions as compared to others.

The remaining chapter is organized as follow: Sect. 2 presents the related work done on energy efficient task offloading techniques. The problem description is presented in Sect. 3. Section 4 described the multi-objective optimization approach. Section 5 explains the proposed modified multi-objective PSO. Section 6 discusses the simulation strategy and result analysis. Finally, Sect. 7 concludes the chapter.

2 Related Works

The various heuristics in the literature have been proposed for task scheduling and task offloading problems in mobile cloud environment. Broadly, these are of two types: (I) minimizing the makespan of an application [16–18] and (ii) minimizing the battery consumption of mobile devices [15, 19, 20]. With the objective of minimizing the makespan, a list based heuristic, HEFT[16] was proposed. Firstly, it assigned priority to all tasks and then mapped the highest priority task to a machine that gave the earlier finish time of a task at each step and thus minimized the overall completion time of an application. Another heuristic, *named*, Push-Pull algorithm has also been proposed. This algorithm initially used a random schedule and then deterministic guided search method is applied to iteratively improve the current solution [17]. For maximizing the throughput a genetic algorithm was proposed [18] that partition the application task over a mobile device and the cloud in an optimized manner. An incremental greedy strategy to offload and parallel execution of perceptual applications [21] has also been proposed to reduce the finish time of applications. In recent years, the main focus of the researchers is on energy aware scheduling mobile devices. Rong and Pedram [22] used the positive slack time between tasks for minimizing the energy consumption of a computer system. Li et al. [23] presented an optimized maximum-flow/minimum-cut task partitioning algorithm to offload the tasks from mobile device to cloud for minimizing energy consumption. Kumar and Lu [24] proposed a strategy based upon computation-to communication ratio for making offloading decision to minimize the energy consumption.

To find the trade-off solutions between completion time and energy for parallel tasks, Lee and Zomaya [25] proposed two energy-conscious scheduling (ECS and ECS + idle) for heterogeneous computing systems. An integer liner programming based optimization technique [26] for adaptive computation offloading is addressed considering the available memory, CPU and energy consumption as the main criteria for offloading. Wu et al. [27] presented an offloading decision model using network unavailability to decide whether to offload a task for remote execution or not. Similarly, another task offloading technique, CRoSS algorithm [28] was also presented using the link failure rate and the bidirectional transmission rate as main factors for offloading. Along with these factors, other important computing factor is clock frequency. Using Dynamic Voltage Frequency Scale (DVFS), the mobile device energy can be further optimized [29] as the CPU clock frequency is approximately linearly proportional to the voltage supply. Similarly, a mobile device can be connected to more than one wireless networks and thus can offload the data to different networks. To address multisite offloading, a graph partitioning approach is proposed to find solution to the partitioning problem [30]. Lin et al. proposed the task scheduling and task migration algorithm from mobile device to cloud based on DVFS for mobile cloud computing [31]. The results showed a significant reduction in energy under the application completion constraints. Energy efficient computational offloading framework (EECOF) [32] has been presented to leverage minimal application processing migration to cloud and thus reducing the total energy consumption cost. Based on contextual network conditions [33], an energy model was presented whether to offload a task or to run it locally. Similarly, few fuzzy and artificial intelligence based decision support systems [34–36] have also been developed to offload the tasks.

From the review of literature, it has been found that most of the existing studies try to minimizing the makespan or energy consumed while scheduling the tasks in mobile cloud environment. None of the existing techniques used multi-objective computational intelligence techniques like MOPSO [37], NSGA-II [38], and FDPSO [39] etc. that give set of near optimal solutions. Hence, this chapter presented multi-objective optimization technique that generates a set of near optimal solutions for mobile cloud applications. We proposed the Modified Multi-Objective Particle Swarm Optimization (MMOPSO) algorithm using the concept of non-dominance to offload the tasks from mobile device to the cloud so as to minimize the energy consumption of created schedule plan.

3   System Model and Assumptions

3.1   Application Model and Mobile Cloud Model

A user application is modelled by a Directed Acyclic Graph (DAG), defined by a tuple G (T, E), where T is the set of n tasks $\{t_1, t_2, ..., t_n\}$, and E is a set of e edges, represent the dependencies. Each $t_i \ \varepsilon \ T$, represents a task in the application and each edge $(t_i ... \ t_j) \ \varepsilon \ E$ represents a precedence constraint, such that the execution of $t_j \ \varepsilon \ T$ cannot be started before $t_i \ \varepsilon \ T$ finishes its execution [40]. If $(t_i, t_j) \ \varepsilon \ T$, then t_i is the parent of t_j, and t_j is the child of t_i. A task with no parent is known as an *entry* task and a task with no children is known as *exit* task. The task size (z_i) is expressed in *Million of Instructions (MI)*.

Our mobile cloud model consists of a mobile device having m, computational cores, $R = \{r_1, r_2, ..., r_m\}$ at different processing power and a cloud resource. The processing power of a core (mobile core or cloud core), is expressed as *Million of Instruction per Second (MIPS) and is denoted by* PP_{r_p}. Each core is Dynamic Voltage Scaling (DVS) enabled; in other words, it can operate with different

Voltage Scaling Levels (VSLs) i.e., at different clock frequencies. This mobile device has also access to the computing resources on the cloud. Each task can be executed on different cores or can be offloaded to cloud for its execution. The execution time, $ET_{(i,p)}$, of a task t_i on a core (either mobile core or cloud core), is calculated by the following equation:

$$ET_{(i,p)} = \frac{Z_i}{PP_{r_p}} \tag{1}$$

We use $ET_{(i,c)}$ to denote the execution time of task t_i on cloud c. Time for sending the task t_i to cloud is given by

$$T_s^i = \frac{data_i}{BW_s} \tag{2}$$

where $data_i$ is the task data and BW_s is the available bandwidth of sending channel. Similarly, time for receiving output of task t_i from cloud is given by

$$T_r^i = \frac{data_i}{BW_r} \tag{3}$$

where $data_i$ is the task data and BW_r is the available bandwidth of receiving channel.

Let $EST\,(t_i, r_p)$ and $EFT\,(t_i, r_p)$ denote the earliest Earliest Start Time and the Earliest Finish Time of a task t_i on a local core r_p, *respectively*. For the entry task, we have:

$$EST\left(t_{entry}, r_p\right) = avail\left(r_p\right) \tag{4}$$

For the other tasks in DAG, we computer EST and EFT recursively as follows:

$$EST\left(t_i, r_p\right) = max \begin{cases} avail\left(r_p\right) \\ \underset{t_j \epsilon pred(t_i)}{max} \left\{AFT\left(t_j\right) + ct_{ij}\right\} \end{cases} \tag{5}$$

$$EFT\left(t_i, r_p\right) = ET_{(i,p)} + EST\left(t_i, r_p\right) \tag{6}$$

where pred (t_i) is the set of parent tasks of task t_i, and *avail* (r_p) is the time when the core r_p is ready for task execution. The Estimated Remote Execution Time of a task t_i on a cloud is given by:

$$ERT(t_i, c) = ET_{(i,c)} + T_s^i + T_r^i \tag{7}$$

Similarly, $AST\,(t_i, r_p)$ and $AFT\,(t_i, r_p)$ denotes the Actual Start Time and Actual Finish Time of task t_i on local core or on cloud, *respectively*. The makespan is equal to the maximum of actual finish time of the exit tasks t_{exit} and is defined by

$$M = max\{AFT(t_{exit})\} \tag{8}$$

The makespan is also referred to as the running time for the entire application DAG. The energy model used in this study is derived from the capacitive power (P_c) of Complementary Metal-Oxide Semiconductor (CMOS)-based logic circuits [41] which is given by:

$$P_c = ACV^2f \tag{9}$$

where A is the number of switches per clock cycle, C is the total capacitance load, V is the supply voltage, and f is the frequency. It's clear from Eq. (9) that the supply voltage is the dominant factor; hence, low supply voltage means lower power consumption.

The energy consumed by executing entire application tasks over available local core is defined as [41]

$$E_l = ACV^2f \cdot ET_{(i,p)} = \alpha V_i^2 ET_{(i,p)} \tag{10}$$

where V_i is the supply voltage of the core on which task n_i is executed, and $ET_{(i,p)}$ is the execution time of task n_i on the scheduled core r_p.

If task n_i is offloaded to the cloud, the energy consumption of mobile device for offloading the task is given by:

$$E_c = ACV^2f \cdot ET_{(i,c)} = \alpha V_i^2 ET_{(i,c)} \tag{11}$$

where V_i is the supply voltage of the sending channel and $ET_{(i,c)}$ is the execution time of task n_i on the cloud c. Therefore, the total energy consumed, i.e., E_{total} for executing the whole application is given by

$$E_{total} = \sum_{i=1}^{n} E_i \tag{12}$$

where $E_i = E_l$ if the task t_i is executed locally and is equal to E_c if the task t_i is offloaded to the cloud.

4 Task Scheduling Based on Multi-objective Particle Swarm Optimization

The first part of this section introduces the concept of Multi-Objective Optimization and the second part gives an overview of Particle Swarm Optimization (PSO).

4.1 Multi-objective Optimization

A Multi-objective Optimization Problem (MOP) [42] with m decision variables and n objectives can be formally defined as:

$$Min(y = f(x) = [f_1(x), \ldots, f_n(x)])$$

where $x = (x_1, \ldots, x_m) \in X$ is an m-dimensional decision vector, X is the search space, $y = (y_1, \ldots, y_n) \in Y$ is the objective vector and Y the objective-space.

In general MOP, there is no single optimal solution with regards to all objectives. In such problems, the desired solution is considered to be the set of potential solutions which are optimal for one or more objectives. This set is known as the Pareto optimal set. Some of the Pareto concepts used in MOP are as follows:

(i) *Pareto dominance*. For two decision vectors x_1 and x_2, dominance (denoted by $\prec$) is defined as follows:

$$x_1 \prec x_2 \Leftarrow \Rightarrow \forall_i f_i(x_1) \leq f_i(x_2) \wedge \exists_j(x_1) < f_i(x_2)$$

The decision vector x_1 is said to dominate x_2 if and only if, x_1 is as better as x_2 for all the objectives and x_1 is strictly superior to x_2 in at least one objective.

(ii) *Pareto optimal set*. The Pareto optimal set P_s is the set of all Pareto optimal decision vectors.

$$P_S = \{x_1 \in X, |\nexists x_2 \in : x2 \prec x1\}$$

where the decision vector, x_1, is said to be Pareto optimal when it is not dominated by any other decision vectors, x_2, in the set.

(iii) *Pareto optimal front*. The Pareto optimal front P_F is the image of the Pareto optimal set in the objective space.

$$P_F = \{f(x) = (f_1(x), \ldots, f_n(x)) | x \in P_S\}$$

4.2 Particle Swarm Optimization (PSO)

Particle Swarm Optimization (PSO) is a stochastic optimization technique that operates on the principle of the social behavior of swarms of birds or the schools of fish [43]. In this technique, a swarm of individuals, known as the particles, flow through the swarm space. Each particle represents a candidate solution to the given problem. Each particle is associated with two parameters, *namely,* current position, x_i and current velocity, v_i.

The position of a particle is influenced by the best position visited by it, i.e., its own experience (*pbest*). Along with *pbest*, the second parameter that influences the position is the position of the best particle in its neighborhood, i.e., the experience of neighboring particles (*gbest*). The performance of each particle is measured using a fitness function that varies depending on the optimization problem. During each PSO iteration k, particle i updates its velocity v_i^k and position vector x_i^k as described below [43]:

(a) *Updating Velocity Vector*

$$v_i^{k+1} = \omega v_i^k + c_1 rand_1 * \left(pbest_i - x_i^k\right) + c_2 rand_2 * \left(gbest - x_i^k\right) \tag{13}$$

where ω: inertia weight; c_1: cognitive coefficient based on particle's own experience; c_2: social coefficient based on the swarms experience; $rand_1, rand_2$: Random variables with between (0,1).

The inertia weight, ω, controls the momentum of the particle. Improvement in performance is obtained by decreasing the value of ω linearly from its maximum value, ω_1, to its minimum value, ω_2 [44]. At iteration k, its value, ω_k is obtained as:

$$\omega_k = (\omega_1 - \omega_2)\frac{\mathbf{max\_k} - k}{\mathbf{max\_k}} + \omega_2 \tag{14}$$

Similarly, if c_1 decreases from its maximum value, c_{1max}, to its minimum value, c_{1min}, then more divergence among the particles in the search space can be achieved, while if c_2 increases from its minimum value, c_{2min}, to its maximum value, c_{2max}, then the particles are much closer to the present *gbest*. The following equations are used to find the values of c_{1i} and c_{2i} at iteration k:

$$c_{1i} = (c_{1min} - c_{1max})\frac{k}{\mathbf{max\_k}} + c_{1max} \tag{15}$$

$$c_{2i} = (c_{2max} - c_{2min})\frac{k}{\mathbf{max\_k}} + c_{2min} \tag{16}$$

where *max_k* is the maximum number of iterations and k is the iteration number.

(b) *Updating Position Vector*

$$x_i^{k+1} = x_i^k + v_i^k \tag{17}$$

where x_i^k: position of the particle at kth iteration; $v_{i:}^k$ velocity of the particle at kth iteration.

(c) **Fitness Function**

The fitness function used in proposed MMOPSO is as described in Eq. (18):

$$Fitness = E_{total} \tag{18}$$

The next section describes the proposed algorithm based upon multi-objective PSO.

5 Proposed Work

In order to solve the multi-objective task scheduling problem for mobile cloud environment, we have proposed the Modified Multi-Objective Particle Swarm Optimization (MMOPSO) algorithm based upon non-dominance sorting procedure. The proposed algorithm is consisting of two phases. In the first phase, the initial schedule in created based upon HEFT [16] to minimizes the makespan. Then in the second phase, the schedule created in first phase is fed into the initial population of MMOPSO for minimizing the energy consumption (E). Both of the phases are explained below:

5.1 First Phase: Initial Schedule

For creating the initial schedule, HEFT algorithm is used to schedule tasks over the mobile cores as well as over the cloud cores. For this purpose, first of all, the application tasks are divided into either local task or cloud task. For each task t_i, we defined its minimum completion time over mobiles cores as

$$T_i^{min} = \min_{1 \leq p \leq m} \{ ET_{(i,p)} \} \tag{19}$$

And if $T_i^{min} < ERT(t_i, c)$, then the task t_i will run on mobile cores and is known as *local task*, otherwise it is known *cloud task*.

If a task t_i is cloud task, then its average execution time is given by

$$w_i = ERT(ti, c) \tag{20}$$

Otherwise,

$$w_i = \underset{1 \leq p \leq m}{avg} \{ET_{(i,p)}\} \tag{21}$$

Each task is assigned a priority using upward rank as defined in HEFT and is given by Eq. (22).

$$rank(t_i) = w_i + \underset{t_j \epsilon succ(t_i)}{max} \{rank(t_j)\} \tag{22}$$

where w_i is the average execution time of the task on the different computing resources; $succ(t_i)$ includes all the children tasks of t_i. After assigning the rank to all tasks, initial schedule is generated using HEFT.

5.1.1 An Example

An example workflow with 10 tasks as shown in Fig. 1a is considered to illustrate the working of the first phase. Figure 1b shows the execution time of these tasks on three different available mobile cores. It has been assumed that $T_s^i = 3$, $T_r^i = 1$ and $ET(t_i,c) = 1$ for each task.

After applying Eq. (18), only task t_2 is identified as *cloud* task and rest will be assigned on the mobile cores. Then the rank of all the tasks is calculated using Eq. (21). The order of execution after sorting tasks in descending order of their rank is: t_1, t_3, t_6, t_2, t_4, t_5, t_7, t_8, t_9, *and* t_{10}. Now the tasks are assigned either on local cores or over cloud using HEFT as shown in the Fig. 2.

5.2 Second Phase: MMOPSO Algorithm

The main steps followed in MMOPSO algorithm are described in Fig. 1. The fitness function used is presented by Eq. (18) in Sect. 4.2. MMOPSO algorithm is executed for bi-objective task offloading problem, i.e., minimization of execution time and energy. Therefore the task offloading problem is formulated as (Fig. 3):

Minimize Time$(S) = max\{AFT(t_{exit})\}$
Minimize Energy $(S) = E_{total}$
Subject to Time $(S) < D$

Where D is the maximum completion time of an application over mobile device. MMOPSO algorithm used the following operators:

(a) *Archive Updating:*

In multi-objective algorithms, the non-dominated particles are stored in elite archive. Particle's dominance is checked against other particles based upon the

Fig. 1 An example

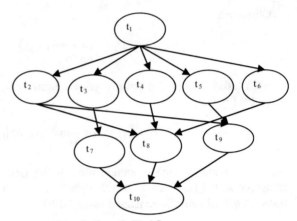

(a)   An Example Workflow

Task	Core1	Core2	Core3
t_1	9	7	4
t_2	8	6	5
t_3	6	5	4
t_4	7	5	3
t_5	5	4	2
t_6	7	6	4
t_7	8	5	3
t_8	6	4	2
t_9	5	3	2
t_{10}	7	4	2

b) **Execution Time of tasks on different cores**

objective functions. The current generation's solutions are combined with the solutions in the archive of previous generations to make *2 N* solutions, where *N* is the size of archive. Then, all of these solutions are sorted in ascending order of their dominance. If more than one solutions show the same dominance value, then diversity perimeter, *I* (.) is calculates for such solutions. The solution showing higher value of *I* (.) is selected. For updating archive, the best *N* solutions are selected from these 2 *N* solutions based upon dominance and perimeter [39].

(b) *Diversity Perimeter*:

The diversity parameter for any solution *y*, *I(y)* is given by:

$$I(y) = \sum_{i=1}^{M} \frac{f_i(x) - f_i(z)}{max(f_i) - min(f_i)} \tag{23}$$

Fig. 2 Initial assignment of first phase

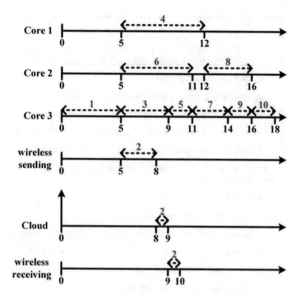

where x and z are adjacent solutions to y, after sorting the solutions in ascending order according to ith objective. The infinite value is assigned to the boundary solutions. Higher the value of $I(y)$, more is sparseness. So, the diversity of the solutions increases with the high values of $I(y)$.

(c) *Updating pbest and gbest:*

The binary tournament operator is used to select *gbest* solution from the current archive. The particle's current position is compared based on dominance sort with the best position from the previous generation for updating *pbest*. If there is no dominate solution, then the current position of particle is selected as current *pbest*.

(d) *Mutation:*

MMOPSO algorithm used the replacement mutation [45] to avoid stucking into local minima and to explore the search space efficiently. For applying the adaptive mutation, mutation probability, *P(Mutation)* is calculated using the following equation:

$$P(Mutation) = 1 - \frac{k}{max\_k} \qquad (24)$$

where k is the current iteration and max_k is the maximum iterations. A random number (*rand*) in range [0, 1] is generated for every particle. If $rand < P$ (*Mutation*), then a task is randomly selected for mutation.

Algorithm : Modified Multi-Objective Particle Swarm Optimization(MMOPSO) algorithm
Input: Application with maximum completion time constraint, D **Output :** Non-dominant energy efficient schedule solutions
1. a) Set iteration counter $k = 0$.   b) Randomly initialize population of N swarm particles.   c) Insert the schedule created in first phase as one of the swarm particle.   d) Initialize all particle velocities $V^k_{(i,)}$ to zeros and personal best position $pbest^k_{(i,)}$ is set to $X^k_{(i,)}$.   2. Evaluate the fitness of all swarm particles according to eq. (18).   3. Based on non-dominance and diversity parameter, sort all the particles in ascending order Then, initialize the archive $A^k_{(i,)}$ with it.   4. Set $k = k + 1$.   5. For all the particles, repeat the following: a) Initialize the $gbest^k_{(i,)}$ from the archive with binary tournament selection.   b) Update the velocity of kth particle $V^k_{(i,)}$ according to Eq. (13).   c) Update particle position $X^k_{(i)}$ according to Eq. (17).   d) Apply adaptive mutation using Eq. (24)   6. Evaluate the fitness of all swarm particles according to eq. (17).   7. Combine the solutions of current particle positions with the archive solutions to have total of $2N$ particle solutions.   8. Select the best N solutions from the solutions of step 7 on the basis of non-dominance sort and perimeter, acc. to Eq. (23) and update the archive.   9. Update each particles $pbest^k_{(i)}$ and $gbest^k_{(i)}$.   10. If ($k <$ max _k) then go to step 3, otherwise output the non- dominant solutions from the archive

Fig. 3 MMOPSO Algorithm

6   Performance Evaluations

In this section, the simulation of the proposed heuristic, MMOPSO is presented. To evaluate the proposed task offloading workflow scheduling algorithm, we used five synthetic workflows based on realistic workflows from diverse scientific applications, which are:

- Montage: Astronomy
- EpiGenomics: Biology
- CyberShake: Earthquake
- LIGO: Gravitational physics
- SIPHT: Biology

The detailed characterization for each workflow including their structure, data and computational requirements can be found in [46]. Figure 4 shows the approximate structure of each workflow.

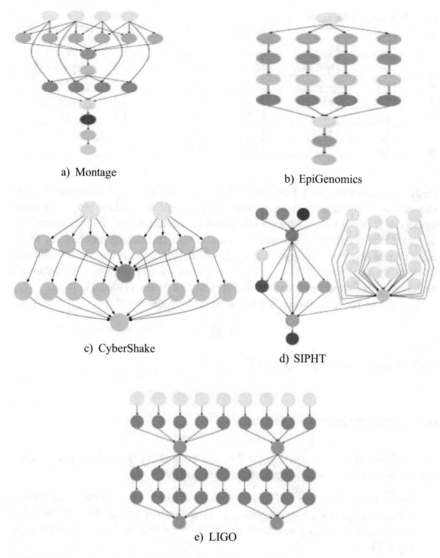

a) Montage

b) EpiGenomics

c) CyberShake

d) SIPHT

e) LIGO

Fig. 4 Structure of various workflows [46]

6.1 Experimental Setup

For simulation, we assume a mobile cloud environment consisting of a mobile device and a cloud service provider. We assume three heterogeneous cores with different processing speed in a mobile device and one core available at the cloud. For simplicity it is assumed that every task take 30 secs to sending data over cloud from the mobile device and take 10 secs to receive the data from the cloud. For this

Table 1 Voltage–relative speed pairs

Level	Pair 1		Pair 2		Pair 3	
	Voltage (v_i)	Relative speed (%)	Voltage (v_i)	Relative speed (%)	Voltage (v_i)	Relative speed (%)
0	1.6	100	2.5	100	2.0	100
1	1.4	85	2.0	75	1.6	80
2	1.2	60	1.5	55	1.4	60
3	1.1	45	1.0	35	1.0	40
4	1.0	30	1.0	35	–	–

study, we have used the CloudSim [47] library. The existing CloudSim simulator allows modelling and simulating cloud environment by dealing only with single workload. It is not suitable for mobile cloud environment. So, the core framework of CloudSim simulator is extended to handle task scheduling problem for MCC. Each core is Dynamic Voltage Scaling (DVS) enabled; i.e., it can work at different Voltage Scaling Levels (VSLs). For each resource, a set V_j of v VSLs is random and uniformly distributed among three different sets of VSLs (Table 1).

The values for maximum completion time, D is generated as:

$$D = 3*M_{HEFT}$$

where $M_{HEFT} = makespan\ of\ HEFT$

6.2   Performance Metrics

The analysis of the proposed algorithm has been done with existing state-of-art algorithms using the following performance metrics:

(a) Generational Distance (GD): GD [38] is a convergence metric and used to access the quality of an algorithm against the true pareto front $P*$ which is generated by merging solutions of different algorithms. It is calculated using Eq. (25):

$$GD = \frac{\left(\sum_{i=1}^{|Q|} d_i^2\right)^{1/2}}{|Q|} \tag{25}$$

where d_i is the Euclidean distance between the solution of Q and the nearest solution of $P*$. Q is the front obtained from algorithms for which GD metric is calculated.

(b) Spacing: To check the diversity among the solutions, spacing metric [38] is used and is given by Eq. (26):

$$Spacing = \sqrt{\frac{1}{|Q|}\sum_{i=1}^{|Q|}(d_i - \bar{d})^2} \tag{26}$$

where d_i is the distance between the solution and its nearest solution of Q and it is different from Euclidean distance and $\bar{d}$ is the mean value of the distance measures d_i. The small value of both GD and Spacing metric is desirable for an evolutionary algorithm.

6.3   *Simulation Results*

This section presents simulation results and analysis of our proposed Multi-objective MMOPSO algorithm. Now a days, Non-dominated Sort Genetic Algorithm (NSGA-II) [38] and ε-FDPSO [39] are the state-of- art techniques to solve MOP. To measure the effectiveness of proposed MMOPSO algorithm, all these algorithms have been designed and simulated for multi-objective workflow scheduling problem for mobile cloud environment. For implementing the NSGA-II, we used binary tournament selection, one-point crossover and replacing mutation. We have assumed parameters used in ε-FDPSO, and MMOPSO algorithms to be: population size = 20, $c_1 = 2.5 \rightarrow 0.5$ and $c_2 = 0.5 \rightarrow 2.5$, inertia weight $\omega = 0.9 \rightarrow 0.1$ and for NSGA-II population size is 20, crossover rate is 0.8, and mutation rate is 0.5. The performance of scheduling algorithms is evaluated considering the randomly generated workflow applications. For bi-objective task offloading problem, we considered the application completion time and the energy consumed of the created schedule as two conflicting objectives. To obtain the Pareto optimal solutions with ε-FDPSO, MMOPSO, and NSGA-II, algorithms, 100 samples have been captured through simulation.

Figures 5, 6, 7, 8 and 9 shows the bi-objectives non-dominated solutions for Montage, CyberShake, EpiGenomics, LIGO, and SIPHT workflows, *respectively*. The x-axis represents the execution time of created schedule for respective workflow structure, and y-axis represents the energy consumed by created schedule for respective workflow structure.

It has been observed that most of the solutions obtained using MMOPSO algorithm is lying closely to the true front and showing he uniform spacing among the solutions. These results are analyzed using two metrics, i.e., GD and Spacing. Table 2 and Table 3 presents the comparative results for all the three algorithms on the basis of GD and Spacing metrics for Montage, CyberShake, EpiGenomics,

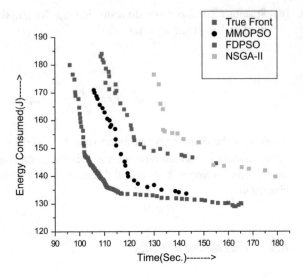

Fig. 5 Bi-objective non-dominated solutions for montage workflow

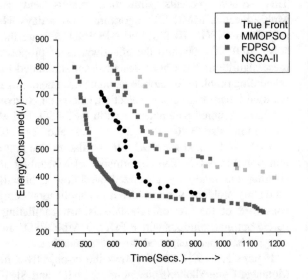

Fig. 6 Bi-objective non-dominated solutions for CyberShake workflow

LIGO, and SIPHT workflows, *respectively.* The results are obtained by taking the average of 10 simulations as described below.

From Table 2, it is clear the performance of the MMOPSO algorithm is better and reaches a solution set that is 84, 83, 55, 55, and 80% closer to true Pareto front as in comparison to the solution set created by FDPSO for Montage, CyberShake, Epigenomics, LIGO and SIPHT workflows, *respectively* as well as 90, 86, 70, 72, and 87% closer to true Pareto front as in comparison to the solution set created by NSGA-II for Montage, CyberShake, Epigenomics, LIGO and SIPHT workflows, *respectively.*

Fig. 7 Bi-objective
non-dominated solutions for
epigenomics workflow

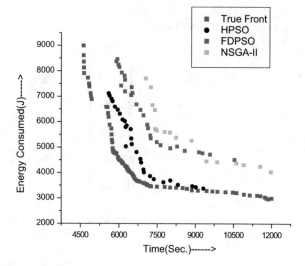

Fig. 8 Bi-objective
non-dominated solutions for
LIGO workflow

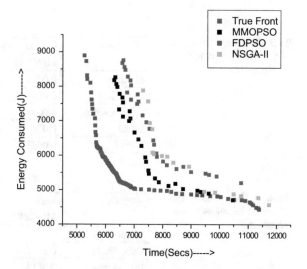

It has been observed from Table 3, the values of spacing metric using MMOPSO algorithm is 26, 43, 38, 36, and 40% lower than that of values of spacing metric obtained using FDPSO for Montage, CyberShake, Epigenomics, LIGO and SIPHT workflows, *respectively* as well as 35, 47, 34, 38, and 37% lower than that of values of spacing metric obtained using NSGA-II algorithm for Montage, CyberShake, Epigenomics, LIGO and SIPHT workflows, *respectively*. This is due to use of trade-off schedule plan between makespan and energy in the creation of non-dominated solution set. So, it is concluded that MMOHPSO algorithm provides uniform spacing as well as better convergence among the solution set as

Fig. 9 Bi-objective
non-dominated solutions for
SIPHT workflow

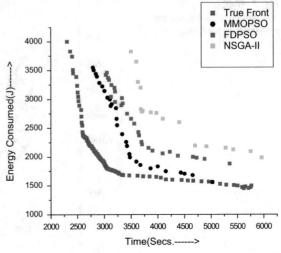

(a) Bi-objective Non-dominated Solutions for SIPHT

Table 2 Comparative results
of GD for all workflow
structures

Workflow	Generational distance (GD)		
	MMOPSO	NSGA-II	FDPSO
Montage	0.0016	0.1016	0.0544
CyberShake	0.0094	0.0396	0.0595
EpiGenomics	0.0097	0.0709	0.037
LIGO	0.0048	0.0358	0.0204
SIPHT	0.0047	0.0423	0.0206

Table 3 Comparative results
of Spread for all workflow
structures

Workflow	Spread		
	MMOPSO	NSGA-II	FDPSO
Montage	0.4736	0.7738	0.6033
CyberShake	0.2157	0.7819	0.7288
EpiGenomics	0.3612	0.5715	0.5426
LIGO	0.285	0.7041	0.6965
SIPHT	0.4315	0.7409	0.7128

compared to other algorithms for all workflow structures under consideration.
Hence, it is applicable to offload large workflows task like face detection and matrix
multiplication etc., over Mobile Cloud environment.

7 Conclusion and Future Work

In the past few years, a single objective task offloading problem has been addresses by many researchers. However, in real life applications, there are multiple conflicting objectives that must be satisfied simultaneously. So, the goal of decision maker is multi-fold and prefers the set of Pareto optimal solutions. To address this issue, we proposed the *Modified Multi-Objective Particle Swarm Optimization* (*MMOPSO*) algorithm based on the concept of dominance to solve the mobile cloud task scheduling problem. It is a combination of multi-objective particle swarm optimization algorithm and list based heuristic. Its performance is analyzed using two conflicting objectives of makespan, and energy consumption under application completion constraints. The efficacy and applicability of the proposed approaches are demonstrated by using different application task graphs and comparing it with state-of-art MOO techniques. The simulation experiments exhibit that MMOPSO performs better and generates the solutions that are more converged towards the true Pareto optimal front and shown uniform spacing among the created solutions. Hence, it is applicable to solve a wide class of multi-objective optimization problems for scheduling tasks over Mobile Cloud environment.

In future, the concept of neural networks, fuzzy logic, etc. needs to be tested for possible enhancement to the proposed heuristic for real life case studies.

References

1. R. Buyya, R. Buyya, C.S. Yeo, C.S. Yeo, S. Venugopal, S. Venugopal et al., Cloud computing and emerging IT platforms: Vision, hype, and reality for delivering computing as the 5th utility. Futur. Gener. Comput. Syst. **25**, 17 (2009). doi:10.1016/j.future.2008.12.001
2. C.J.R. Gabriel, M., Wolfgang G., Hybrid computing—where hpc meets grid and cloud computing. J Futur. Gener. Comput. Syst. **27**, 440–453 (2011)
3. S. Abrishami, M. Naghibzadeh, Deadline-constrained workflow scheduling in software as a service cloud. Sci. Iran. **19**, 680–689 (2012). doi:10.1016/j.scient.2011.11.047
4. R. Van Den Bossche, K. Vanmechelen, J. Broeckhove, Online cost-efficient scheduling of deadline-constrained workloads on hybrid clouds. Futur. Gener. Comput. Syst. **29**, 973–985 (2013). doi:10.1016/j.future.2012.12.012
5. S. Abrishami, M. Naghibzadeh, D.H.J. Epema, Deadline-constrained workflow scheduling algorithms for Infrastructure as a Service Clouds. Futur. Gener. Comput. Syst. **29**, 158–169 (2013). doi:10.1016/j.future.2012.05.004
6. P. Parwekar, From Internet of Things towards cloud of things, in *2nd International Conference on Computer and Communication Technologies ICCCT-2011*, 2011, pp. 329–333. doi:10.1109/ICCCT.2011.6075156
7. A. Botta, W. De Donato, V. Persico, A. Pescape, On the integration of cloud computing and internet of things, in *International Conference on Future Internet Things Cloud, FiCloud 2014*, 2014, pp. 23–30. doi:10.1109/FiCloud.2014.14
8. A. Botta, W. De Donato, V. Persico, A. Pescapé, Integration of Cloud computing and Internet of Things: A survey. Futur. Gener. Comput. Syst. **56**, 684–700 (2016). doi:10.1016/j.future.2015.09.021

9. D. Gachet, M. De Buenaga, F. Aparicio, V. Padrón, Integrating internet of things and cloud computing for health services provisioning: the virtual cloud carer project, in *6th International Conference on Innovative Mobile Internet Services Ubiquitous Computing IMIS 2012*, 2012, pp. 918–921. doi:10.1109/IMIS.2012.25

10. R. Petrolo, V. Loscrí, N. Mitton, Towards a smart city based on cloud of things, in *ACM International Workshop Wireless Mobile Technologies Smart Cities—WiMobCity '14*, 2014, pp. 61–66. doi:10.1145/2633661.2633667

11. S.Y. Chen, C.F. Lai, Y.M. Huang, Y.L. Jeng, Intelligent home-appliance recognition over IoT cloud network, in *9th International Wireless Communications on Mobile Computing Conference IWCMC 2013*, 2013, pp. 639–643. doi:10.1109/IWCMC.2013.6583632

12. R.C. Andrea Prati, R. Vezzani, M. Fornaciari, Intelligent video surveillance as a service, in *Intelligent Multimedia Surveillance*, ed. by A.C. Pradeep, K. Atrey, M.S. Kankanhalli (Springer, Berlin, 2013), pp. 1–16

13. M. Yun, B. Yuxin, Research on the architecture and key technology of internet of things (IoT) applied on smart grid, in *International Conference on Energy Science and Electrical Engineering (ICAEE)*, 2010: pp. 69–72. doi:10.1109/ICAEE.2010.5557611

14. B.B.P. Rao, P. Saluia, N. Sharma, A. Mittal, S.V. Sharma, Cloud computing for Internet of Things & sensing based applications, in *Sixth International Conference on Sensing Technology (ICST)* 2012, pp. 374–380. doi:10.1109/ICSensT.2012.6461705

15. C.M. Sarathchandra Magurawalage, K. Yang, L. Hu, J. Zhang, Energy-efficient and network-aware offloading algorithm for mobile cloud computing. Comput. Netw. **74**, 22–33 (2014). doi:10.1016/j.comnet.2014.06.020

16. M.Y. Wu, T Haluk, H. Salim, Performance-effective and low-complexity task scheduling for heterogenous computing. IEEE Trans. Parallel Distrib. Syst. **13**, 260–274 (2002)

17. S.C. Kim, S. Lee, J. Hahm, Push-pull: deterministic search-based DAG scheduling for heterogeneous cluster systems. IEEE Trans. Parallel Distrib. Syst. **18**, 1489–1502 (2007). doi:10.1109/TPDS.2007.1106

18. L. Yang, J. Cao, S. Tang, T. Li, A.T.S. Chan, A framework for partitioning and execution of data stream applications in mobile cloud computing, in *IEEE 5th International Conference* on *Cloud Computing CLOUD 2012*, 2012, pp. 794–802. doi:10.1109/CLOUD.2012.97

19. E. Ilavarasan, R. Manoharan, High performance and energy efficient task scheduling algorithm for heterogeneous mobile computing system. J. Comput. Sci. **2**, 10–27 (2010)

20. Z. Tang, L. Qi, Z. Cheng, K. Li, S.U. Khan, K. Li, An energy-efficient task scheduling algorithm in DVFS-enabled cloud environment. J. Grid Comput. **14**, 55–74 (2016). doi:10.1007/s10723-015-9334-y

21. M. Ra, A. Sheth, L. Mummert, P. Pillai, D. Wetherall, R. Govindan, Odessa : enabling interactive perception applications on mobile devices categories and subject descriptors, in *Proceedings of the International Conference* on *Mobile Systems, Applications*, and *Service*, 2011, pp. 43–56. doi:10.1145/1999995.2000000

22. P. Rong, M. Pedram, Power-aware scheduling and dynamic voltage setting for tasks running on a hard real-time system, in *Asia and South Pacific Design Automation Conference*, 2006, pp. 473–478. doi:10.1109/ASPDAC.2006.1594730

23. Z. Li, C. Wang, R. Xu, Task allocation for distributed multimedia processing on wirelessly networked handheld devices, in *16th IEEE International Parallel & Distributed Processing Symposium*, 2001, pp. 741–746. doi:10.1109/IPDPS.2002.1015589

24. K. Kumar, Y. Lu, Cloud computing for mobile users : computation save energy ? Computer (Long. Beach. Calif) **43**, 51–56 (2010)

25. Y.C. Lee, A.Y. Zomaya, A novel state transition method for metaheuristic-based scheduling in heterogeneous computing systems. IEEE Trans. Parallel Distrib. Syst. **19**, 1215–1223 (2008). doi:10.1109/TPDS.2007.70815

26. D. Kovachev, T. Yu, R. Klamma, Adaptive computation offloading from mobile devices into the cloud, in *10th IEEE International Symposium on Parallel and Distributed Processing and Applications ISPA 2012*, 2012: pp. 784–791. doi:10.1109/ISPA.2012.115

27. W. Huijun, H. Dijiang, S. Bouzefrane, Making offloading decisions resistant to network unavailability for mobile cloud collaboration, in: *International Conference on Collaborative Computing: Networking, Applications* and *Worksharing (Collaboratecom), 2013*, pp. 168–177. doi:10.4108/icst.collaboratecom.2013.254106
28. S. Ou, K. Yang, L. Hu, CRoSS: a combined routing and surrogate selection algorithm for pervasive service offloading in mobile ad hoc environments, in *GLOBECOM—IEEE Global Telecommunications Conference, 2007*, pp. 720–725. doi:10.1109/GLOCOM.2007.140
29. L.H.L. Huang, Q.X.Q. Xu, Energy-efficient task allocation and scheduling for multi-mode MPSoCs under lifetime reliability constraint,in *Design, Automation & Test* in *Europe Conference & Exhibition, (DATE), 2010*. doi:10.1109/DATE.2010.5457063
30. K. Sinha, M. Kulkarni, Techniques for fine-grained, multi-site computation offloading, in *IEEE/ACM International Symposium on Cluster, Cloud,* and *Grid Computing CCGrid 2011*, 2011, pp. 184–194. doi:10.1109/CCGrid.2011.69
31. X. Lin, Y. Wang, Q. Xie, M. Pedram, Task scheduling with dynamic voltage and frequency scaling for energy minimization in the mobile cloud computing environment. IEEE Trans. Serv. Comput. **8**, 175–186 (2015). doi:10.1109/TSC.2014.2381227
32. M. Shiraz, A. Gani, A. Shamim, S. Khan, R.W. Ahmad, Energy efficient computational offloading framework for mobile cloud computing. J. Grid Comput. **13**, 1–18 (2015). doi:10.1007/s10723-014-9323-6
33. M. Akram, A. Elnahas, Energy-aware offloading technique for Mobile cloud computing, in *Proceedings of the 2015 International Conference on Future Internet* of *Things and Cloud, FiCloud 2015. International Conference* on *Open* and *Big Data, OBD 2015*, 2015, pp. 349–356. doi:10.1109/FiCloud.2015.45
34. O.W. Samuel, G.M. Asogbon, A.K. Sangaiah, P. Fang, G. Li, An integrated decision support system based on ANN and Fuzzy_AHP for heart failure risk prediction. Expert Syst. Appl. Elsevier Publishers **68**, 163–172 (2017)
35. A.K. Sangaiah, A.K. Thangavelu, X.Z. Gao, N. Anbazhagan, M.S. Durai, An ANFIS approach for evaluation of team-level service climate in GSD projects using Taguchi-genetic learning algorithm. Appl. Soft Comput. **30**, 628–635 (2015)
36. A.K. Sangaiah, A.K. Thangavelu, An adaptive neuro-fuzzy approach to evaluation of team-level service climate in GSD projects. Neural Comput. Appl. Springer Publishers **23**(8) (2013). doi:10.1007/s00521-013-1521-9
37. C.C. Coello, G.T. Pulido, M.S. Lechuga, Handling multiple objectives with particle swarm optimization. IEEE Trans. Evol. Comput. **8**, 256–279 (2004). doi:10.1109/TEVC.2004.826067
38. K. Deb, A. Pratap, S. Agarwal, T. Meyarivan, A fast and elitist multiobjective genetic algorithm: NSGA-II. IEEE Trans. Evol. Comput. **6**, 182–197 (2002). doi:10.1109/4235.996017
39. R. Garg, A.K. Singh, Multi-objective workflow grid scheduling using ε-fuzzy dominance sort based discrete particle swarm optimization. J. Supercomput. **68**, 709–732 (2014). doi:10.1007/s11227-013-1059-8
40. A. Verma, S. Kaushal, Cost-time efficient scheduling plan for executing workflows in the cloud. J. Grid Comput. **13**, 495–506 (2015). doi:10.1007/s10723-015-9344-9
41. M. Mezmaz, N. Melab, Y. Kessaci, Y.C. Lee, E.G. Talbi, a. Y. Zomaya, et al., A parallel bi-objective hybrid metaheuristic for energy-aware scheduling for cloud computing systems, J. Parallel Distrib. Comput. **71**, 1497–1508 (2011). doi:10.1016/j.jpdc.2011.04.007
42. K. Deb, *Multi-Objective Optimization Using Evolutionary Algorithms*(Wiley, 2005), pp. 13–46
43. J. Kennedy, R. Eberhart, Particle swarm optimization, in *Proceedings of the IEEE International Conference on Neural Networks, 1995*, vol. 4, pp. 1942–1948. doi:10.1109/ICNN.1995.488968
44. P.K. Tripathi, S. Bandyopadhyay, S.K. Pal, Multi-Objective Particle Swarm Optimization with time variant inertia and acceleration coefficients. Inf. Sci. (NY) **177**, 5033–5049 (2007). doi:10.1016/j.ins.2007.06.018

45. R. Garg, Multi-objective optimization to workflow grid scheduling using reference point based evolutionary algorithm. Int. J. Comput. Appl. **22**, 44–49 (2011)
46. S. Bharathi, A. Chervenak, E. Deelman, G. Mehta, S Lanitchi, K. Vahi, et al., Characterization of scientific workflows, in *Workflows in Support of Large-Scale Science, CA, USA, 2008*, pp. 1–10. doi:10.1109/WORKS.2008.4723958
47. R.N. Calheiros, R. Ranjan, A. Beloglazov, C.A.F. De Rose, R. Buyya, CloudSim: a toolkit for modeling and simulation of cloud computing environments and evaluation of resource provisioning algorithms. Softw. Pract. Exp. **41**, 23–50 (2011). doi:10.1002/spe.995

Distributed Algorithm with Inherent Intelligence for Multi-cloud Resource Provisioning

Seyed Ali Miraftabzadeh, Paul Rad and Mo Jamshidi

Abstract Dynamic distributed algorithm for provisioning of resources has been proposed to support heterogeneous multi-cloud environment. Multi-cloud infrastructure heterogeneity implies the presence of more diverse sets of resources and constraints that aggravate competition among providers. Sigmoidal and logarithmic functions have been used as the utility functions to meet the indicated constraints in the Service Level Agreement (SLA). Spot instances as the elastic tasks can be supported with Logarithmic functions while the algorithm always guaranteed Sigmoidal functions have the priority over the Logarithmic functions. The model uses diverse sets of resources scheduled in a multi-clouds environment by the proposed Ranked Method (RM) in a time window "slice". To maximize the revenue and diminish cost of services in the pooled aggregated resources of multi-cloud environment, the multi-dimensional self-optimization problem in distributed autonomic computing systems is proposed.

Keywords Cloud computing · Scheduler · Multi-clouds · Federated cloud · Spot instances

This work was supported, in part, by Open Cloud Institute at University of Texas at San Antonio, Texas, USA and by Grant number FA8750-15-2-0116 from Air Force Research Laboratory and OSD, USA. The authors gratefully acknowledge use of the services of Chameleon cloud and Jetstream cloud, funded by NSF awards 1419165 and 1445604 respectively.

S.A. Miraftabzadeh (✉) · P. Rad (✉)
Open Cloud Institute, The University of Texas, San Antonio, TX, USA
e-mail: Ali.Miraftab@utsa.edu

P. Rad
e-mail: Paul.Rad@utsa.edu

S.A. Miraftabzadeh · P. Rad · M. Jamshidi
ACE Laboratories, The University of Texas, San Antonio, USA
e-mail: moj@wacong.org

© Springer International Publishing AG 2017
A.K. Sangaiah et al. (eds.), *Intelligent Decision Support Systems for Sustainable Computing*, Studies in Computational Intelligence 705, DOI 10.1007/978-3-319-53153-3_5

1 Introduction

According to a forecast from International Data Corporation (IDC) [1] the world-wide spending on public cloud services is expected to surpass $107 billion in 2017 [2]. Among different forms of delivering cloud services, IDC recognized the Infrastructure as a Service (IaaS) model as one of the fastest growing categories with compound annual growth rate of 27.2%. IaaS is a promising solution for enabling on-demand access to an elastic pool of configurable and virtual computational services (e.g., computing power, storage, and networks) in a pay-as-you-go manner. IaaS providers offer computational services in the form of Virtual Machine (VM) with specific resource characteristics such as computing power, memory, disk space and networking bandwidth along with types of operating systems and installed applications. Despite traditional distributed computing systems such as Grid [3], cloud computing focuses on the monetary focal point while promising flexibility and high performance for users and cost-efficiency for operators. Furthermore, taking advantage of the virtualization technology gains more resource utilization and Return On Investment (ROI) [4].

In addition, cloud computing is increasing its penetration among industry, research institutes, and universities across applications. An increasing amount of computing is now performed on public clouds, such as Amazon's EC2, Windows Azure, Rackspace, and Google Compute Engine [5–7] or on private clouds hosted by enterprise using open source OpenStack [8] software. In addition, numbers of cloud computing research test beds and production sites such as Chameleon or JetStream have been supported by National Science Foundation (NSF) to support academic research [9].

As a drastic increment of the customers, a variety of requested types and emerging cloud computing providers profit maximization for customers and providers becomes a more entangled problem in terms of minimizing the cost using resource management [4, 10–14], performance improvement in networking and applications [15, 16], maximizing request revenue by providing multi price services [17], market segmentation [18] and prognostication the demands [19].

Considering the natural perishable characteristic of the cloud resources produces a primary conclusion to maximize the resource utilization and to avoid wasting the non-storable cloud resources. On the other hand, honoring the associated Service Level Agreement (SLA) which guarantees the certain level of Quality of Service (QoS) imposes another robust constraint and increases the complexity of the problem. A topic of recent interest, the federated cloud and multi-cloud deployment, allows different cloud providers to share resources for increased scalability and reliability. The aim of the Multi-Clouds is to integrate resources from different providers in a way that access to the resources is transparent to users and cost can be reduced [20]. Furthermore, the Multi-Clouds is a possible mechanism for maximizing a provider's profit by leasing part of the available resources in the pool computing center for the requested bids by other providers [20–22]. So the

Multi-Clouds implicitly maximizes the resource utilization while keeping the required QoS.

Maximizing the resource utilization requires the resource allocation decision which involves determining what, how many, where, and when to make the resource available to the user [23]. Typically, users choose the type and amount of the cloud resources and providers place the petition for specific ones onto nodes in their datacenter. Running the application efficiently entails matching the type of the resources and the workload characteristics. In addition, the amount should be sufficient to meet the indicated constraints in SLA. In a pay-as-you-go environment like the cloud where users can request or return resources dynamically, time scheduling such as an adjustment is also important to consider. To maximize resources' utilization, the job scheduler is responsible for allocating desired resources to an individual job and in parallel it should be guaranteed that each job is given an adequate amount of resources, or its fair share. Such a scheduling decision becomes more complex in the cloud computing as a heterogeneous environment with two players: cloud providers and cloud users. On one side, there are the users who have fluctuating loads with a variety of applications and requests. On the other side, consolidated cloud environments are likely to be constructed from a variety of machine classes, representing different options in configuring computing power, memory and storage [24]. The Multi-Clouds is an exceedingly multi-parts hetero-geneous system as a result of sharing diversity sets of datacenters in terms of technical performance and monetary strategies.

Mathematical interpretation of effectively allocating datacenters' resources in the Multi-Clouds is a multi-dimensional self-optimization problem in distributed autonomic computing systems. As a bridge between providers and users, utility functions have very smart theoretical properties and their use in practical autonomic computing systems started to explore in [25, 26]. Utility functions can enable a collection of autonomic elements to continually optimize the use of computational resources in a dynamic, heterogeneous environment [27].

There are a lot of works that have been done for two major cloud platforms' advantages: producing the flexible available resources and minimizing the cost of the datacenters [28–30]. Nevertheless, relatively less work has been done on the provider's sideways to maximize revenue and diminish cost of services produced. In this paper, based on the Sigmoidal and Logarithmic utility functions the pro-posed algorithm guaranteed the dynamic and scalable job scheduling in the heterogeneous Multi-Clouds environment for the most flourishing cloud service, IaaS. In parallel, the algorithm guaranteed the minimal management effort or ser-vice providers' interaction. The algorithm can rapidly provide and release Multi-Clouds pool resources at once period of processing decision.

2 Background in Resource Management and Upcoming Challenge

This section briefly review resource management strategies founded in grid and cloud computing, covering topics such as the compute model, data model, virtualization, monitoring, and provenance. These topics are extremely important to understand the main challenges that both grids and cloud face today, and will have to overcome in the close future [31].

Usually grid computing usesa batch-scheduled compute model in which a local resource manager (LRM) administers the compute resources for a Grid site, and users submit batch jobs (via GRAM) to request some resources for the specific period of time. For instance, PBS, Condor, and SGE are common LRM. Grid computing is not efficient for interactive applications; one of the most obvious reasons for this inefficiently is grid has policies enforcing the system to identify the users and credentials under which the job will run for. However, cloud computing model gets the advantageous of sharing the resources by all users at the same time; in contrast to dedicated grid resources strictly controlled by a queuing system in grid computing. This approach overcome the problem for serving the latency sensitive applications; however, the downside is ensuring a good enough level of QoS. As the cloud grows in scale and number of users, this problem turns to the dominant challenge.

3 Ranked System in Multi-cloud

Two main stakeholders in cloud computing, providers which are serving cloud services and customers which are client or consumers of the produced services, have their own inevitable momentous operative influence to make the cloud environment as a heterogeneous system. Particularly, heterogeneous cloud computing datacenter is considered in this section to overcome the resource management problems in multi-cloud.

Collaboration across the clouds requires strapping monitoring and management mechanism which are the key components for provisioning, scheduling and failure management [32]. Efficiently provisioning computing resources in a pool of heterogeneous datacenters has been done by the proposing Ranking Method. Producing a dynamic sequence of the available resources based on the hierarchical characteristics of the cloud providers is the first step of the proposed algorithm. The dynamic sequence could be matched with dynamic pricing which it has been acknowledged by the literature [33, 20]. Since then the model could be considered as one of the dominant methods to maximize the request revenue and consequently get the most out of the resource utilization. The dynamic pricing can be explored by using intermediate parameter as Bids between customers and cloud providers. In addition, the Ranking Method can be administrated based on most instantaneous

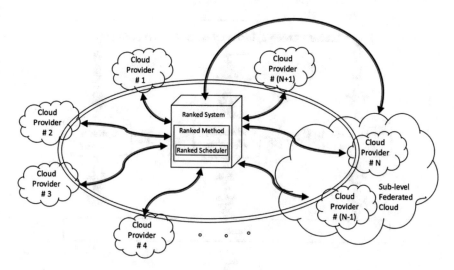

Fig. 1 Illustration of ranking method

parameters, in [34] more detail for effective parameters are described. Figure 1 illustrates the high level topology of the Ranked system. The Ranked System can have been embedded on cloud embodiment or it can function as an independent scheduler system. It consists of Ranked Method and Ranked Method consists of Ranked Scheduler. The Ranked system communicates not only to cloud providers directly but also to other types of cloud topologies such as Multi-Clouds directly or in directly and monitor and control ongoing processes.

Distributed scheduling system consists four major components, Fig. 2. Following the API, user interface communicates with the Ranked System distributed in Configuration Database, Scheduling Algorithm, and Operation Database. Configuration Database feeds the Scheduling Algorithm with the assigned utility functions. Using intelligent forecasting, utility functions describe the requested services while considering the providers' concerns; for instance, maximizing the resource allocation while minimizing the power consumption. Operation Database applies the final decision of the Scheduling Algorithm and inspects online activities of consumers and providers; the aim of the latter task is for real time notification for the Scheduling Algorithm.

Ranked System follows the following clues:

- similar instances of different providers are situated in the same group (which is named clutch)
- initially top ranked instances have the uppermost slots and bottom ranked instances have the lowermost slots
- for customer satisfaction, evaluation of the requested service is placed in a specific group based on the lowest and highest requested instances; interpolation and other methods can be used

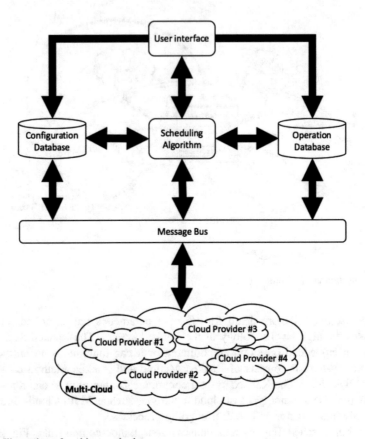

Fig. 2 Illustration of ranking method

- before fully occupied resources, top ranked instances are allocated to the corresponding request
- after fully occupied resources, top ranked instances may not be allocated for the request but another ranked instance in a same clutch has more opportunity to be allocated to the task. This action strictly depends on the assigned utility function to the request.
- transcendent ranked instances ameliorate customer satisfactions for the assigned service
- transcendent ranked instances are unintended to be occupied if the number of demand increases
- inferior ranked instances keep the request revenue in order to maximize the cloud revenue
- instances in a same clutch can be reshuffled to maximize the resource utilization if there is profit for the corresponding providers

- hedging from clutch to the adjacent clutch is possible in a case of resource shrinking and strictly depends on the utility function
- each clutch could include different providers

Details of Ranked system architecture are depicted in Fig. 3. Web Application and API is the gate for managing applications, services, clouds; it supports Rest based API alongside any other types of input which can be done through User Interface. Configurable management is the manager of models, assemblies, environment; it matches the demands from the consumers to the user profile. Ranked Method is responsible for creating scheduling design and deploy the design; more explanation is provided in the following paragraph. Heuristic Evaluation which is responsible of comparing deployed design to planned design, consumers' activities and consumers' updates. Messenger Dispatcher provides the details monitoring facilities for User Interface with the metrics data. Elastic-Search Service records all deployments, thread, dataflow, task mapping, assemblies, and environments. The attached engine to it provides log analytics, real-time application monitoring, and click stream analytics. Publisher tracks the Ranked Method and post the design on the massage bus. Auxiliary Service helps implementing policies and costs models.

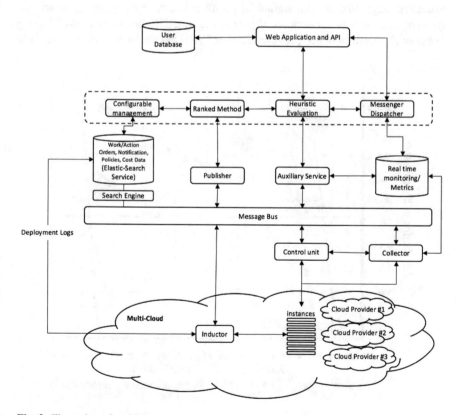

Fig. 3 Illustration of ranking system

Real time storage of the metrics data which are provided by Collector, is a database having the features such as scalability and high availability without compromising performance. Internal communication between the components is carried on by Message bus. Controller is responsible for distributing action orders. Functionality of the collector includes metrics data from instances for any event. multi-cloud contains an Inductor and the pool of shared. Inductor receives the design and deploy the actions and post the results message to the system. It is responsible for executing steps in implementation, gather the queue of clouds specifications, and make the reaction queue and report it to controller.

3.1 Utility Functions in Ranked System

In this section a scenario is studied to clarify how the proposed algorithm gets advantage of the sigmoidal utility function; the concept can be extended to logarithmic function which we use for elastic tasks. Figure 4 illustrates four clutches in which five providers make a pool of ranked instances as small, medium, large and extensive large. Six unalike sigmoidal utility functions corresponding to the six types of customers or services are shown in Fig 4. Utility functions 1 and 2 belong to the small clutch in which provider number 5 (P5) has the highest priority to serve

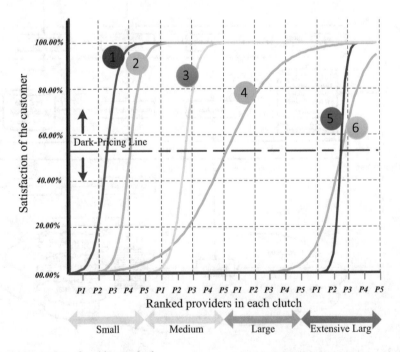

Fig. 4 Illustration of ranking method

any request in the clutch. In this clutch customers who are assigned to utility function 2, rather than 1, has more chance to take the better service based on the chosen ranked parameter.

However, utility function 1 gets services simultaneously from P5 if enough resources are available in the Multi-Clouds. By increasing demands and reducing the number of free instances, the chance of getting services from inferior provider increases. The Ranking Method tries to provision resources for all of types of the customers. However, by increasing the number of requests for a specific provider or clutch, based on the Service Level Agreement (SLA), three scenarios are likely to be happened. First, pass the customer to the lowermost neighbor provider or lower ranked provider. Chosen desired sigmoidal function by the customers carries on this scenario automatically. Second, the customer may lose allocated instances if it does not satisfy the provider proposed conditions such as raising the price. This scenario can be implemented by control the whole number of available resources, $(V - z)$, in each processing period. Finally, the customer accepts the proposed condition and the allocated instances are remained to serve the customer. Utility function 3 describes the only service in the medium clutch. It spread over the whole clutch from P1 to P5; so each of the serving providers can provision the customer. For large clutch only utility function 4 is presenting services but it spreads not only in large clutch but also from small clutch to extensive large clutch. Consequently, customers in this utility function have a potential to get serve in a variety clutches; the customers have a chance to have an experience in extensive large clutch when enough pooled resources are available. Apparently, it is not a case when Multi-Clouds has an experience in high amount request. This types of utility function expansively helps a lot to maximize the revenue of the requests by pro-viding motivation such as paying less for this types of functions. Finally, in extensive clutch two types of utility functions are managing the requests revenue and resources of the providers. As it is illustrated utility function 5 has the sharpest slop in this simplified properties illustration of Ranking Method. This types of utility function can be fixed in a specific provider with two applications: first, serves a customer by a specific ranked or no ranked properties over a time. Next, fixed the instance of the provider for a service after the customer accepts the dynamic conditions of the provider such as dynamic pricing. Figure 5 shows how clutches are formed in the Ranked Method. 801 delivers the specifications of the cloud providers to 802 to categorize the instances based on VDs properties, for example Computational power, Memory Size, and Hard Drive capacity. To this point it is refer to collecting the information. After the collecting process the selection process is established. 804 is refer to the weighting phase; 803 observes the dynamic and static parameters and informs 804. Dynamic parameters could be any variables which may vary; such as measurement unit, occupation percentage of a cloud provider, networking parameters, elastic resources. Static parameter such as: service type, security, Quality of Service (QoS), pricing unit, instances size, operating system, pricing sensitivity to regions, subcontractor, and many other. 805 shows the Indexing steps. It is based on statistics and predictions. Predicting the variation is not limited to the cloud providers side, it is connected to the user data also. Indexing

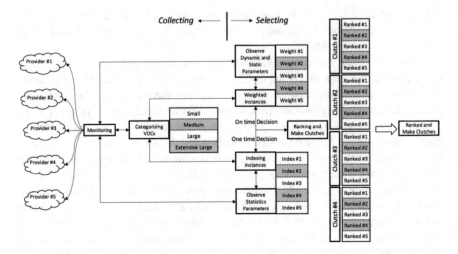

Fig. 5 Illustration of ranking method

steps uses the activity and history of the cloud providers and customers to predict the workload. Machine Learning algorithms and other methods can be used in this steps. The weighting phase monitors the system continuously. However, indexing phase is one-time decision; it is carried on in every specific period of time. 807 makes the ranked instances in a queue using the result of weighting and indexing phases.

So far, it is described that how the Ranking Method provides flexibility to serve the customers dynamically, by applying some principle conditions in the SLA to match the customer requests and utility functions. Maximizing the profits of the cloud providers, as one of the primary goals of the Multi-Clouds, makes the Ranking Method as a method to potentiate the dynamic pricing conditions in the SLA. In the next step mathematical formulation would be studied and the Ranking Method is presented as the distributed algorithm in the Multi-Clouds.

4 Mathematical Formulation of the Algorithm

In this section, we will present design and implementation of the distributed algorithm of optimized controller for resource allocation of elastic and inelastic traffic using logarithmic and sigmoidal-like utility functions. This algorithm is based on the utility proportional policy to maximize the requests revenue where the fairness among users is in utility percentage and describes the user gratification with the service of the corresponding bid. Furthermore, this algorithm precisely provides information of the available resources and distributes them among the users. It could be implemented as a part of the cloud service brokerage or as a separate unit in cloud computing architecture of IaaS.

Mathematical formulation of the algorithm includes *Sigmoidal* [35] and *Logarithmic* [36] utility functions which have had applications in resource scheduling. These are used as gauges for dynamically resources allocation such that predefined required constrains by SLA are satisfied and cloud providers get the benefit of using all their available resources and minimizes the operating costs. Virtual data center (vDC) is considered as the resource unit; however, the approach can straightforwardly be extended to consider a variety kinds of instances which include different amount of compute, memory, storage and bandwidth resources.

The vDC allocated by controller to the kth service is given by v_k. Presumption of the algorithm is suitable utility function is assigned to the service by using techniques such as interpolation, statistic methods or others. Our objective is to determine vDC that the scheduler should allocate to the specific customer. For this optimization problem we are looking for the strictly concave utility functions that satisfies the following properties:

(a) $U_k(v_k)$ is an increasing function of v_k.
(b) $U(0) = 0$ and $U(\infty) = 1$.
(c) $U_k(v_k)$ is twice continuously differentiable in v_k.
(d) U_k is bounded above.

Note that typically, most utility functions in current networks can be represented by three types of functions:

- Sigmoidal-like
- Strictly concave function
- Strictly convex function

Next, let us consider two major type of tasks: Sigmoidal-like Tasks and Logarithmic Tasks.

Sigmoidal-like Tasks

First type refers to the task that requires the specific amount of resources which are requested by the service. The characteristic of the requested *pay-as-you-go* service match to the suitable sigmoidal function. Usually, inelastic tasks can be defined by sigmoidal function. It can be expressed as:

$$U_k(v_k) = c_k \left(\frac{1}{1 + e^{-a_k(v_k - b_k)}} - d_k \right) \tag{1}$$

to satisfy all the required properties it is enough to have $c_k = \frac{1 + e^{a_k b_k}}{e^{a_k b_k}}$ and $d_k = \frac{1}{1 + e^{a_k b_k}}$. By choices a_k and b_k describe the utility function of the kth user. For example, by choosing $a = 5$ and $b = 10$ a good approximation for a step function can be obtained which explains the fixed usage of the allocated resources; for instance, users with the specific requirement of the resources. Normalized sigmoidal-like utility function with $a = 0.5$ and $b = 20$ is another example for representation of the adaptive real-time application.

Logarithmic Task

It is desired to use all resources in a pool of vDCs as much as possible. In a case of temporary real-time services, the same as spot instances, normalized logarithmic utility function describes the good description. It is referred to the mission of which a minimum number of resources are needed and it has a desire to occupy more resources in a case of availability. Mathematical expression of the function is as:

$$U_k(v_k) = \frac{\log(1 + m_k v_k)}{\log(1 + m_k v_{max})} \tag{2}$$

where v_{max} is the point in which the task occupies all available resources as same as what is mentioned in the SLA. For example, if the user is fully satisfied by 30 vDCs, v_{max} should be set to 30. Any allocated vDCs less than 30 leads to the minus satisfaction but it does not mean user is loss its performance since the performance is based on the required resources for doing the specific job. It means if the user is using all 30 vDCs there is no request issues from the user for share the resources to others. But if the user is not fully using allocated resources and it is confirmed by the monitoring and negotiation process then the user can share the resources by management process. In such a case the customer and the provider get the advantage simultaneously. In (2) m_k is the rate of increasing utility percentage with the allocated rate v_k.

Both function types are satisfied the constraints as in (a) to (d). In Fig. 6. the normalized sigmoidal-like and Logarithmic examples of utility functions are shown.

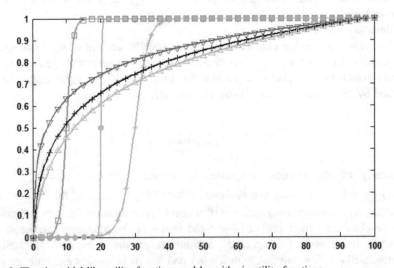

Fig. 6 The sigmoidal-like utility functions and logarithmic utility functions

To maximize the resource allocation based on the utility function the objective function is defined as follow:

$$\max_{v} \prod_{k=1}^{M} U_i(v_i)$$

$$\text{Subject to } \sum_{k=1}^{M} v_k \leq R$$

$$v_k \geq 0 \, i = 1, 2, \ldots, M$$

$$(3)$$

where $v = \{v_1, v_2, v_3, \ldots, v_M\}$ and M is the number of the requests. This formulation ensures non-zero resource provisioning for customers while guarantees minimum predefined constraints are met. Furthermore, using *Sigmoidal* function leads to the more resource allocation for corresponding tasks when the specific amount of resources is required.

Now we do have an optimization problem for proportional resource allocation based on the utility functions [37]. Prove of existence of the global optimal solution for (3) is straightforward. Since (3) is a convex optimization problem a unique trace-back global optimal solution is exists for the problem.

Objective function of $\arg\max_v \prod_{k=1}^{M} U_k(v_k)$ is equivalent to $\arg\max_v \sum_{k=1}^{M} \log(U_k(v_k))$ and for both applied functions, Sigmoidal and Logarithmic

$$\frac{d}{dv_i} \log U_i(v_k) > 0$$

$$\frac{d^2}{dv_i^2} \log U_i(v_k) < 0$$

$$(4)$$

So both functions are strictly concave natural logarithms. Therefore, the optimization problem is a convex optimization problem and there exist the unique trace-back global optimal solution [38].

4.1 Dynamic Distributed Resource Provisioning Approach

To make the optimization problem as the negotiated process for provision resources dynamically, based on demand and supply, this paper gets advantage from the dual problem to distribute the optimization problem among the Multi-Clouds stakeholders. In communication network it has been done similarly in [37, 39, 40] by defining the Lagrangian function as:

$$L(v,p) = \sum_{k=1}^{M} \log(U_k(v_k)) - p\left(\sum_{k=1}^{M} v_k + z - R\right)$$

$$= \sum_{k=1}^{M} (\log(U_i(r_i)) - pv_k) + p(R - z) \tag{5}$$

$$= \sum_{k=1}^{M} L_k(v_k, p) + p(R - z)$$

where $z \geq 0$ is the slack variable (we discuss in more detail about the slack variable later) and p is Lagrange multiplier or the dark price which is the intermediate parameter in negotiation between the Multi-Clouds and customers. To match the ranked instances by Ranking Method which is proposed in the paper, dark price is weighted by the ranked order which is called bid in the paper. Bidding strategy attracts a lot of researcher especially for spot instances [41, 42]. By assigning a set of utility functions not only spot instances but also other tasks can be evaluated simultaneously to make the decision for resources provisioning in the Multi-Clouds.

Satisfying the simultaneous scheduling for multi types of tasks the kth bid for the instance can be given by $w_k = pr_k$. Subsequently, the dual problem objective function can be written as:

$$D(p) = \max_{v} L(v, p) \tag{6}$$

and the dual problem is given by

$$\min_{p} D(p)$$
$$\text{Subject to } p \geq 0 \tag{7}$$

to find the optimal answer driving a derivation of the new optimization problem leads to the interesting formulation for dark price by using $\sum_{k=1}^{M} w_k = p \sum_{k=1}^{M} r_k$ and solving the equation for p:

$$\frac{\partial D(p)}{\partial p} = R - \sum_{k=1}^{M} v_k - z = 0$$
$$p = \frac{\sum_{i=1}^{M} w_i}{R - z} \tag{8}$$

latest equation entails critical information to optimize the problem for p:

- Summation of the bids (from all requests)
- Available Instances (all the active resources)

which both of them are available in the centralized node, monitoring and negotiation process.

On the other hand since the $L(v, p)$ is separable in v we can rewrite the equation as:

$$\max_{v} \sum_{k=1}^{M} (\log(U_k(r_k)) - pv_k) = \sum_{k=1}^{M} \max_{r_k}(\log(U_k(r_k)) - pv_k) \qquad (9)$$

which implies, the optimization problem can be solved for each utility function separately:

$$\max_{v_k}(\log(U_k(v_k)) - pv_k)$$
$$\text{Subject to } p \geq 0 \qquad (10)$$
$$v_k \geq 0 \, i = 1, 2, \ldots, M$$

Equations (7) and (10) distribute the optimization problem (3) into two parts. Equation (7) is in quest of the dark price which requires supply and demands information and can be considered to be solved in the Multi-Clouds. Equation (10) which separately maximizes provisioning of the resources for each utility function and it can be counted up as a customer's attorney. Between these two optimization problems dark price is an intermediate parameter to moderate the criteria for both sides. Establish an iterative process to drawn out the settle down point in which stakeholders get the mutual advantage is the chosen approach of the algorithm.

The mutual proliferation leads to the maximization of the request revenue and utilization of the pooled vDCs in Multi-Clouds environment.

5   Proposed Algorithm for Resource Scheduling

So far, the reasons and requirements for implementation of the expedient scheduling algorithm have been studied in various aspects. Putting together all the described aspects of the proposed algorithm has been projected in two mutually joint algorithms, Back shown in Algorithm (1) and Feed in Algorithm (2). In this section terms explanation and mathematical formulation of the algorithms have been described. Dark Price is considered as the iterative parameter between Back-Algorithm which is the customers' attorney and Feed-Algorithm which is the multi-cloud lawyer. Feedback process will continue to settle down in profitable point for customers and providers in the pool of vDCs. Each active tasks send the initial bid to the broker. The broker makes a decision based on the difference of two consequent bids per task with the pre-specified gage as δ. If the absolute difference gets greater than δ the broker specify the new shadow price based on the utility function type. Each user receives the shadow price and solve its own optimization problem $v_{k(n)} = \max_{v_k}(\log(U_k(v_k)) - pv_k)$ for v_k which is used to calculate the new bid $w_k(n) = p(n)r_k(n)$. The broker collects all the bids sequentially and the process would be repeated until $|w_k(n) - w_k(n-1)|$ gets less than the pre-specified threshold δ. The Back and Feed are the same as UE and eNodeB algorithm in [37]. There are modifications which are applied in the algorithm to match them with Ranked Method.

Algorithm 1 Back (k^{th} Utility Function)	**Algorithm 2** Feed
Initiate its bid as $w_k(1)$ to Feed	**loop**
loop	input $V - z$ and scale the functions by $(V-z)^{-1}$
Received shadow price $p(n)$ from the Feed	if k^{th} utility function is the active
if STOP from Feed **then**	Receive bids $w_k(n)$ from Back-Algorithm {Let $w_k(0) =$
Calculate allocated rate $v_n^{opt} = \frac{w_k(n)}{p(n)}$	$0 \, \forall_i$}
STOP	**if** $\|w_k(n) - w_k(n-1)\| < \delta \, \forall_k$ **then**
else	Allocate rates, $v_k^{opt} = \frac{w_k(n)}{p(n)}$ to k^{th} Utility Function
Solve $v_k(n) = \max_{v_k}(\log(U_k(v_k)) - pv_k)$	STOP
Send new bid $w_k(n) = p(n)r_k(n)$ to Feed-Algorithm	**else**
end if	Based on the specified group for p_k:
end loop	Calculate $p(n) = \frac{\sum_{k=1}^{M} w_k^{\Box}(n)}{V-z}$
	Send new shadow price based on the users category
	end if
	end loop

Scaling factor as (V-z) to make the algorithm as an applicable process in the proposed Ranked Method. In the following chapter simulation results declare the convergence of the algorithm in scale which is the requirement of the sustainable computing.

5.1   Simulation Results

Ranking method is applied to the variety sets of utility functions. Convergence of the proposed algorithm is acknowledged by the simulation. In the simulations, to show the generality of the algorithm, it is supposed that instances of the providers are ranked from 1 to 100 which they are included in four clutches with the same size. So one can interpret each number to the instances with the specific parameters. In the first simulation, six utility functions corresponding to sigmoidal functions in Fig. 4. are simulated. Through classification of the instances it is assumed almost the same instances are put to the same clutch and for clearly illustrate of the algorithm only computing power is studied as the gauge to make distinguish between the instances. However, the algorithm has the potential to study all the instances properties such as computing power, memory, disk space and networking bandwidth factors at one simulation. In this case the classification methods should be used to weight the gauge. In the first simulation the computing power is used as the weighted parameter to provide the gauge. The necessitated manipulation to the algorithm is related to the $(V - z)$ as the maximum available resources. The pool of resources in vDC of the Multi-Clouds is the other explanation of the $(V-z)$ manipulation. In the algorithm V is the maximum number of resources and z is the reserved resources for the specific tasks or customers. Since by increasing $(V-z)$

just the scaling of the functions are varied, the convergence of the optimization problem is not denied. The manipulation which is used for this simulation is as follow:

$$(V - z) = \sum_{i=1}^{I} \sum_{n}^{N} C_{i,n} P_i^n \tag{11}$$

Which I is the maximum number of providers, N is the maximum number of the clutches $C_{i,n}$ is coefficient for the ith provider in the nth clutch and P_i^n is the corresponding provider advantageous over other providers in the lucrative specification terms. The scaling factor is $(V - z)^{-1}$ for all the functions. In the first simulation $P_i^n = 1$ which indicates similarity in the performance of the providers and $C_{i,n} = b_{i,n}$.

Figure 7 describes the convergence of the algorithms by weighting the ranked clutches in the x axis and scaling all the functions based on the available calculated resources. $(V - z) = 240$ is considered with the iterations $n = 20$. The rates of different functions with the number of iteration are illustrated in Fig. 7. After giving enough iterations the algorithm is settle down for the sigmoidal functions as 13, 21, 38, 6, 89 and 76 correspondingly. This simulation shows for the 4th sigmoidal function the allocated instances is placed in the lowest clutch.

It means the algorithm provides the lowest priority for the 4th function since the 4th function (with $a = 0.12$) is spread throughout the operating boundaries.

In the second case the available resources are decreased to $(V - z) = 100$. From the results in Fig. 8 and referring to Fig. 4 it can be understood sigmoidal function number 5 is satisfied with the appropriate instances which is matched with its request. However sigmoidal function number 6 is not confined in its primary clutch and it is downgraded to the small clutch. The reason for that is sigmoidal function

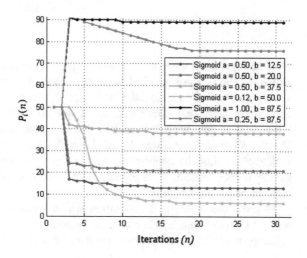

Fig. 7 Allocated ranked providers convergence $P_i(n)$ with number of iterations n for $(V - z) = 240$

Fig. 8 Allocated ranked
providers convergence $P_i(n)$
with number of iterations n
for $(V - z) = 100$

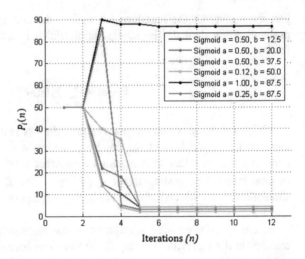

Fig. 9 Allocation of ranked
providers for
$100 \le (V - z) \le 300$

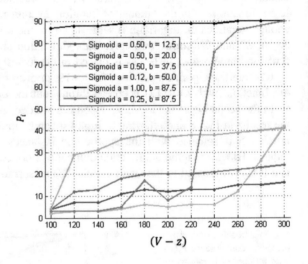

number 6 has the *a* value as 0.25 which spreads the function outside of its primary
clutch despite of the sigmoidal function number 5 with the *a* value as 1 which is
strictly confined the sigmoidal function in its primary clutch. The same reason is
applied for function number 4. Since it spreads widely throughout the operating
boundaries it gets even less ranked instance than function number 1, 2 and 3.
Finally, sigmoidal functions 1 and 2 get the same instances which means low
ranked clutch has the opportunely to serve more customers. It means in a case of
having higher requests than the available resources, petitions for the low ranked
clutches arise dramatically.

In multi-cloud this is the situation which competition between the providers
arises to serve more number of customers by offering lower size of instances.
Actually when there are enough available resources the algorithm provides the

customers by the highest performance resources based on their desired utility functions. However, when the requests are getting higher than the available resources, based on the customers chosen model, customers with bounded utility functions maintain their primary resources with the small variation but the allocated resources for other types of functions are varied to satisfy each chosen model by the customers. In Fig. 9 The steady state rate of different sigmoidal functions with different $(V - z)$ has been illustrated. As mentioned before monitoring system in parallel with brokering strategies control the situation and decide the final decision. Study the utility function number 4 and 6 is interesting. Since the requested resources by this functions have not been satisfied until the rest functions reach their corresponding requested resources.

5.2 Implementation of Spot Instances with the Logarithmic Function

The distributed algorithm has the interesting application in spot instances pricing model to control the resource allocation. Spot instances is following the method which impose dynamic price based on supply and demand. Since maximizing revenue of requests is one of the Multi-Clouds aims, a dynamic resource allocation has been simulated.

Figure 10 shows the sigmoidal and logarithmic functions with their character-istics which are used as reference for Figs. 11 and 12. Figure 11 illustrates the distributed algorithm gives the sigmoidal functions higher priority than the

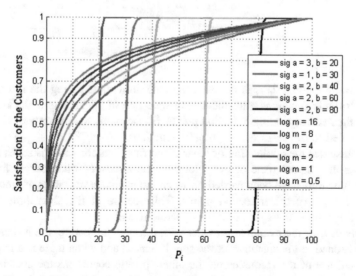

Fig. 10 The sigmoidal and the logarithmic utility functions

Fig. 11 Convergence of the
allocation for the ranked
providers $P_i(n)$ for sigmoidal
and logarithmic functions for
$(V-z) = 500$

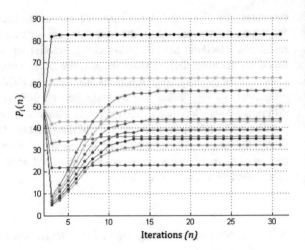

Fig. 12 Allocation of ranked
providers for sigmoidal and
logarithmic functions for
$220 \leq (V-z) \leq 500$

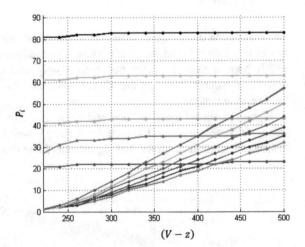

logarithmic function. This is because, the algorithm starts giving the resources to
the logarithmic functions after the steady state rate of all the sigmoidal functions
exceed their corresponding inflection points. Furthermore, the majority of resources
are allocated to the tasks with sigmoidal functions.

To illustrates the application of logarithmic functions for spot instances Fig. 12
describes the characteristic of the algorithm; the allocation of resources for logarithmic functions have been carried on when the available resources are more than
the summation of the inflection points. This feature of the algorithms has the
potential to use as the spot instances dynamic model.

In the case of the available provider with the specific feature, the instances can
be provisioned to the customer of the spot instances and when there is a request of
permanent use of the resources on the other specific conditions the spot is return
back to the system as the available resource.

6 Conclusion

The Ranking Method has been introduced as the elastic and inelastic tasks scheduler to support the heterogeneous requests and resources in multi-cloud environment. The convergence of the algorithm is studied by simulation for a variety sets of the sigmoidal and logarithmic functions while computations are spread over two mutually joint algorithms, Back algorithm (customers' attorney) and Feed algorithm (multi-cloud's lawyer). Sigmoidal function has been used to make a balance between requests and resources while always SLA requirements are guaranteed. To maximize the usage revenue, in a pool of resources, the combination of logarithmic and sigmoidal functions has been proposed to support the spot instances for the elastic tasks while guaranteed the sigmoidal functions always have the priority; which means the algorithm support the elastic and inelastic applications simultaneously. In addition, demonstration of the Ranked method elucidates the capabilities to handle the dynamic resource provisioning in scale in multi-cloud system. The substantial Ranked system is proposed to support the Ranked method in the heterogeneous multi-cloud environment. One of the main limitation in scheduling algorithm is the delay response of the system which is strongly suggested for the future research for the Ranked system.

References

1. International Data Corporation. http://www.idc.com/
2. A.N. Toosi, On the economics of infrastructure as a service cloud providers: pricing, markets, and profit maximization (2014)
3. I. Foster, C. Kesselman, *The Grid 2: Blueprint for a New Computing Infrastructure* (Elsevier, 2003)
4. A. Beloglazov, R. Buyya, Managing overloaded hosts for dynamic consolidation of virtual machines in cloud data centers under quality of service constraints. IEEE Trans. Parallel Distrib. Syst. **24**(7), 1366–1379 (2013)
5. Google Compute Engine. https://www.cloud.google.com/products/compute-engine/
6. Amazon EC2. http://www.aws.amazon.com/ec2/
7. Windows Azure. http://www.azure.microsoft.com/
8. Openstack cloud software. http://www.openstack.org/
9. Chameleoncloud. https://www.chameleoncloud.org/
10. P. Rad, V. Lindberg, J. Prevost, W. Zhang, M. Jamshidi, ZeroVM: secure distributed processing for big data analytics, pp. 1–6
11. D. Hancock, C. Stewart, J. Fischer, J. Lowe, P. Rad, M. Vaughn, *Resource Management from HPC to the Cloud: Do you Manage Resources or do they Manage you?* (2016)
12. P. Rad, A. Chronopoulos, P. Lama, P. Madduri, C. Loader, Benchmarking bare metal cloud servers for HPC applications, in *2015 IEEE International Conference on Cloud Computing in Emerging Markets (CCEM)*, pp. 153–159 (2015)
13. S.M. Balakrishnan, A.K. Sangaiah, MIFIM—Middleware solution for service centric anomaly in future internet models. *Future Generation Computer Systems* (Elsevier Publishers, 2016). doi:10.1016/j.future.2016.08.006

14. S.M. Balakrishnan, A.K. Sangaiah, Integrated QoUE and QoS approach for optimal service composition selection in internet of services. *Multimedia Tools Applications* (Springer Publishers, 2016). doi:10.1007/s11042-016-3837-9
15. P. Rad, R. V. Boppana, P. Lama, G. Berman, and M. Jamshidi, Low-latency software defined network for high performance clouds, pp. 486–491
16. M. Muppidi, P. Rad, S.S. Agaian, M. Jamshidi, Container based parallelization for faster and reliable image segmentation, pp. 1–6
17. A. Gohad, N.C. Narendra, P. Ramachandran, *Cloud Pricing Models: A Survey and Position Paper*, pp. 1–8
18. W. Wang, B. Li, B. Liang, Towards optimal capacity segmentation with hybrid cloud pricing, pp. 425–434
19. L. Zhang, Z. Li, C. Wu, Dynamic resource provisioning in cloud computing: a randomized auction approach, pp. 433–441
20. M. Mihailescu, Y.M. Teo, Dynamic resource pricing on federated clouds, pp. 513–517
21. E. Elmroth, F.G. Márquez, D. Henriksson, D.P. Ferrera, Accounting and billing for federated cloud infrastructures, pp. 268–275
22. B. Rochwerger, D. Breitgand, E. Levy, A. Galis, K. Nagin, I. M. Llorente, R. Montero, Y. Wolfsthal, E. Elmroth, J. Caceres, The reservoir model and architecture for open federated cloud computing. IBM J. Res. Dev. **53**(4), 4: 1–4: 11 (2009)
23. G. Lee, Resource allocation and scheduling in heterogeneous cloud environments: University of California, Berkeley (2012)
24. C. Reiss, A. Tumanov, G.R. Ganger, R.H. Katz, M.A. Kozuch, Heterogeneity and dynamicity of clouds at scale: Google trace analysis, p. 7
25. A. Byde, M. Sallé, C. Bartolini, Market-based resource allocation for utility data centers. *HP Lab, Bristol, Technical Report HPL-2003-188* (2003)
26. T. Kelly, Utility-directed allocation
27. W.E. Walsh, G. Tesauro, J.O. Kephart, R. Das, Utility functions in autonomic systems, pp. 70–77
28. L.A. Barroso, *Warehouse-Scale Computing: Entering the Teenage Decade* (2011)
29. L.A. Barroso, J. Clidaras, U. Hölzle, The datacenter as a computer: An introduction to the design of warehouse-scale machines. Synth. Lect. Comput. Archit. **8**(3), 1–154 (2013)
30. J. Hamilton, Cost of power in large-scale data centers, 11. http://www.perspectives.mvdirona.com/
31. I. Foster, Y. Zhao, I. Raicu, S. Lu, Cloud computing and grid computing 360° compared, pp. 1–10
32. M. Kozuch, M. Ryan, R. Gass, S. Schlosser, D. O'Hallaron, Cloud management challenges and opportunities, pp. 43–48
33. H. Xu, B. Li, Dynamic cloud pricing for revenue maximization. IEEE Trans. Cloud Comput. **1**(2), 158–171 (2013)
34. S. Sundareswaran, A. Squicciarini, D. Lin, A brokerage-based approach for cloud service selection, pp. 558–565
35. J.-W. Lee, R.R. Mazumdar, N.B. Shroff, Downlink power allocation for multi-class wireless systems. IEEE/ACM Trans. Netw. (TON) **13**(4), 854–867 (2005)
36. G. Tychogiorgos, A. Gkelias, K.K. Leung, Utility-proportional fairness in wireless networks, pp. 839–844
37. A. Abdel-Hadi, C. Clancy, A utility proportional fairness approach for resource allocation in 4G-LTE, pp. 1034–1040
38. S. Boyd, L. Vandenberghe, *Convex Optimization*. (Cambridge university press, 2004)
39. S.H. Low, D.E. Lapsley, Optimization flow control—I: basic algorithm and convergence. IEEE/ACM Trans. Netw. (TON) **7**(6), 861–874 (1999)

40. G. Anastasi, E. Borgia, M. Conti, E. Gregori, Rate control in communication networks: shadow prices proportional fairness and stability, J. Cluster Comput **8**(2–3), 135–145 (2005)
41. Y. Song, M. Zafer, K.-W. Lee, Optimal bidding in spot instance market. pp. 190–198
42. S. Karunakaran, R. Sundarraj, *Bidding Strategies for Spot Instances in Cloud Computing Markets* (2014)

Parameter Optimization Methods Based on Computational Intelligence Techniques in Context of Sustainable Computing

Pankaj Upadhyay and Jitender Kumar Chhabra

Abstract In sustainable computing techniques, we always need to solve lots of optimization problems like design, planning and control, which are extremely hard. Conventional mathematical optimization techniques are computationally difficult. Recent advances in computational intelligence have resulted in an increasing number of nature inspired metaheuristic optimization techniques for effectively solve these complex problems. Mainly, the algorithms which are based on the principle of natural biological evolution and/or collective behavior of swarm have shown a promising performance and are becoming more and more popular nowadays. Most of these algorithms have their some set of parameters. The performance of these algorithm is highly depends on optimal parameter value settings. Prior to running these algorithms, the user must have values of different parameters, such as population size, parameters related to selection, and crossover probability, number of generations etc. That is energy and resource consuming. In this paper we summarize the work in computational intelligence based parameter setting techniques, and discuss related methodological issues. Further we discuss how parameter tuning affects the performance and/or robustness of metaheuristic algorithms and also discusses parameter tuning taxonomy.

Keywords Sustainable computing · Nature inspired metaheuristic · Parameter optimization · Meta-optimization

P. Upadhyay (✉) · J.K. Chhabra
Computer Engineering Department, NIT Kurukshetra, Kurukshetra, Haryana, India
e-mail: pankajupadhyay2006@gmail.com

J.K. Chhabra
e-mail: jitenderchhabra@gmail.com

© Springer International Publishing AG 2017
A.K. Sangaiah et al. (eds.), *Intelligent Decision Support Systems for Sustainable Computing*, Studies in Computational Intelligence 705, DOI 10.1007/978-3-319-53153-3_6

101

1 Introduction

Sustainable computing is concerned with computational methods for sustainable economy, environment and society. It is a broad filed which attempts to optimize computational energy and resource utilization using techniques from mathematical optimization and computer science field. For example Intelligent Transportation System (ITS) is used to provide maximum comfort and convenience to its commuters and further minimizes the operating cost, energy consumption and green house emission. It requires lot of optimization process. In sustainable computing we always require to solve lot of optimization problems like design, planning and control, which are computationally hard. In many problems conventional mathematical optimization methods are computationally difficult. Recent advances in computational intelligence have been resulted in an increasing number of nature inspired metaheuristic optimization technique for effectively solve these complex problems [1].

In literature there are various algorithms which are based on nature inspired computing to solve optimization problems, and some of them such as particle swarm optimization (PSO), grey wolf, ant colony optimization (ACO), Artificial bee colony (ABC), genetic algorithm (GA)and gravitational search algorithm (GSA), have been given very prominent results.

Swarm intelligence and nature inspired metaheuristics are most efficient and widely used algorithms for optimization purpose. These algorithms are more efficient over conventional mathematical optimization algorithms [2, 3]. Every optimization algorithm has some strength and some weakness over other optimization algorithms, some work well on certain problem classes while others may not. According to Wolpert and Macready [4], in heuristic search there is no algorithm which gives better results than all other algorithms for all problems. One of the major issues in applying metaheuristics is how to set optimal parameters. Computational complexity is very high to adjust the control parameters of the algorithm to improve its performance on a particular problem. Choosing the optimal parameter values for a single algorithm to solve a single problem is already non-trivial. Parameter setting is an optimization problem itself. Optimal parameter setting is not only affecting the efficiency of actual optimization problem but also improves the performance and robustness of optimization algorithm. All computational intelligence (CI) algorithms are intrinsically dynamic and adaptive process. Hence the uses of fixed parameters that do not change their values during run are against the spirit of dynamism. This implies that performance of algorithm is dependent on whether parameters are static or dynamic in nature. In literature, numerous studies focused on automatic optimal parameter setting. Though various techniques have been proposed in literature, most of them are computationally expensive when the number of parameters is very high.

The objectives of this paper are given as follows:

- To discuss parameter optimization taxonomy.
- To discuss issues and challenges in parameter optimization and why it is so important in computational sustainability.

- To discuss and compare recent computational techniques used for parameter optimization on different performance indicators.

2 Parameter Optimization Taxonomy

Parameter optimization is the process of setting optimal control parameters for optimization technique that is used for optimization problems. Eiben [5, 6], categorizes the adjusting parameters in two categories: parameter tuning and parameter control. The taxonomy for parameter setting is given below in Fig. 1.

- Parameter tuning: Parameter tuning is the approach of finding the best parameter values before starting the algorithm and parameters remain fixed throughout the runtime of the algorithm.
- Parameter control: In parameter control, parameter values are fixed at the beginning but change during execution.

  - Deterministic Parameter Control: This approach is used when algorithm parameter is change by using some deterministic rule. This deterministic rule is used to change the algorithm parameters without using any feedback from the search strategy.
  - Adaptive Parameter Control: This type of parameter control approach is takes place when there is use of feedback to change algorithm parameters.
  - Self-adaptive Parameter Control: In this approach meta-evolution is used to evolve the parameter values during run of baseline algorithm. During evolution better parameter values propagate from one iteration to another.

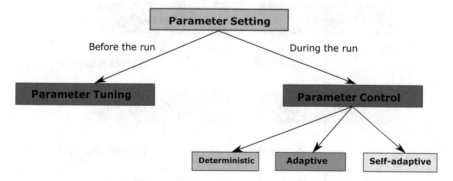

Fig. 1 Taxonomy of parameter setting

3   Parameter Optimization Technique

One of the starting approaches for optimal parameter setting is factorial design, in this technique a comprehensive search is perform to tune the parameter values. This method performs evaluation of the objective function using different parameter values and gets the optimal parameter settings. However, it is time consuming and inefficient so it is often avoided. The cost of such process can be reduced by applying some heuristics, which allow a non-exhaustive search of optimal parameters. Parameter settings are generally chosen in practice by hit and trial method, and tuned by hand [7], taken from other fields, by parameter tuning [8] or by adaptation and self-adaptation mechanisms (parameter control) [9].

In the field of computational intelligence (CI) traditionally there are two main approaches to choose parameter values.

- Parameter tuning, where (good) parameter setting is done before the run of a given CI algorithm. Here, parameter values are remain fixed during CI algorithm is running.
- Parameter control, where (good) parameter setting is fixed during the run of a given CI algorithm. Here, parameter values undergo changes during the run of CI algorithm.

During the last decade there has been extensive research into parameter control. It has been successfully applied in computational intelligence approaches, including Evolution Strategies [10–13], Genetic Algorithms [12, 14], Differential Evolution [15, 16] and Particle Swarm Optimization [17].

For better understanding of underlying parameter tuning approach it is better to further divide it in layered structure. It has three layers namely application layer, algorithm layer and design layer shown in Fig. 2.

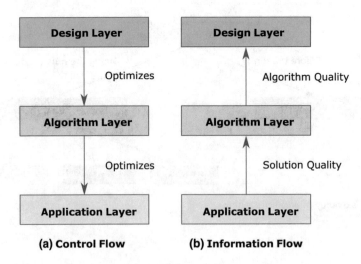

Fig. 2   3 layered architecture of parameter tuning

Table 1 Vocabulary used in parameter optimization

	Problem solving	Parameter tuning
Method	Computational intelligence algorithm	Tuning algorithm
Search space	Solution vector	Parameter vector
Quality	Fitness	Utility
Assessment	Evaluation	Test

The first layer of the architecture contains a design method that is used to find out optimal parameter settings for the underlying computation intelligence technique. This design method can be CI Algorithm or interactive session with user itself. At second layer the algorithm present itself for which parameter setting is to be fixing and at the third layer problem description is available [8]. The vocabulary for distinguishing entities in context of problem solving algorithm and in parameter tuning algorithms is given in Table 1.

The problem solving part contains the underlying CI method to solve the problem where as parameter tuning part contains the metaheuristic method to find optimal parameter setting. The search space for problem is solution vector and for parameter tuning is parameter vector of different values of parameter. Quality check for problem solving method is the fitness function (objective function) and for parameter tuning it is utility.

3.1 Meta-optimization Methods

In the late 1970s by Mercer and Sampson [18] has been given a meta-optimization technique for optimal parameter value settings. It is one of the earliest automatic parameter optimization methods. But due to the large computational costs, their research was very limited. A simple way of finding good behavioral parameters for an optimizer is to employ another overlaying optimizer, called the meta-optimizer. The main concept of meta-optimizer is shown in Fig. 3. In meta-optimization at the lowest level there is optimization problem and at mid level the optimizer to the problem itself. At the top level meta-optimizer is available to solve parameter optimization for problem optimization algorithm.

Another work was done by Grefenstette [19], who also used a GA to optimize the discrete parameters of a GA. He optimized against a set of low dimensional test-function problems to find the generally optimal parameters. In his approach Grefenstette perform extensive experimentation to show the effectiveness of GA as meta-optimizer.

Individuals of population used in meta-optimizer at design layer are parameter vectors of numerical values. Each of the values presented in parameter vector belong to one of the parameter of the underlying CI algorithm to be tuned. To evaluate the utility of parameter vector, the underlying CI algorithm is run several

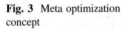

Fig. 3 Meta optimization
concept

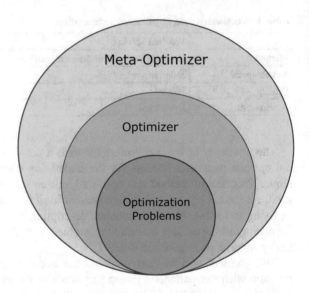

times using these parameter values. Using this approach for representation and performance (fitness) basically any CI algorithm can be used as meta-optimizer.

Bäck [20], optimize parameter settings by using a specialized hybrid of GAs and ES as meta-optimizer. He used first time a parallel master-slave approach to overcome computational limitations. A more advanced meta-optimization method called REVAC has been developed by Nannen and Eiben [21], which is not only able to optimize the parameters, but also to estimate the relevance of each parameter. Recently Pedersen [22] used the local unimodal sampling (LUS) as a meta-optimizer to find the optimal parameters for a differential evolution (DE) algorithm. Since meta-optimization is very time-consuming, it is not feasible to perform realistic experiments. Now computational power increases manifolds at a scale that allowed to performing realistic experiments.

In literature there are various parameter optimization algorithms based on meta-optimizer. Here are some points about meta-optimization methods.

- In most approaches the implementation is strongly coupled with algorithms that are in use. At meta-level mostly variants of existing metaheuristics were used.
- While specialized meta-level algorithms have the advantage that they can be optimized for a low number of evaluations, the drawback is that they are not easily exchangeable by other existing algorithms.
- A very few authors used parallelization concepts for optimization, but most of them only mentioned that it is highly suitable for meta-optimization.
- Most of the approaches have only aim to optimize the base problem, which turns into poor quality of algorithm. Robustness and performance are considered by only few authors.

3.2 Self Tuning

In [23], Yang et al. have been given a self tuning framework for parameter optimization. In this method algorithm it-self is used to tune the parameters. This kind of parameter tuning is highly expensive and very tough to implement. They demonstrated this framework using firefly optimization algorithm. Proposed algorithm simultaneously tunes parameters itself with actual optimization problem. Parameter tuning and optimality finding has been done simultaneously.

For unconstrained standard optimization problem the goal is to find out global minimum f* of a objective function f(x).

$$\text{Minimize } f(x), \quad \text{given } x = (x_1, x_2, x_3, \ldots\ldots, x_d) \tag{1}$$

An algorithm Algo is used to solve this optimization problem to find the fmin value that is within tolerance to global minimum f*. The aim of parameter tuning is to find out best parameter setting P*. Thus the parameter tuning algorithm can be formulated as follows

$$\text{Minimize } Algo(f(x), P), \quad \text{given } P = [P_1, P_2, \ldots\ldots, P_n] \tag{2}$$

In self-tuning framework authors viewed this problem as multi-optimization problem in which two objective functions will be optimized.

$$\text{Minimize } f(x) \text{ and Minimize } Algo(f(x), P) \tag{3}$$

Self tuning algorithm is described as shown in Fig. 4.

There are various approaches to solve multi-objective optimization problem like Pareto optimality, weighted sum. Basically this bi-objective optimization problem can use any of these methods. Here are some concluding remarks about self-tuning parameter optimization.

Implement an algorithm *Algo*(f(x), P, ε) with P=[P₁, P₂, ...Pₙ], ε =[ε₁, ε₂, ... εₖ]
    Define a tolerance
        Algorithm objective (f(x), p, ε) ;
        Problem objective function f(x) ;
        Find the optimality solution f_{min} within tolerance ;
        Output the number of iterations needed to find f_{min} ;
    Solve min (f(x), p) using *Algo* (f(x), p, ε) to get the best parameters;
Output the tuned algorithm with the best parameter setting p∗.

Fig. 4 Self tuning framework

- Optimal parameter setting in any optimization algorithm is highly depends on the very own optimization problem, and there is no unique solution for all problems.
- Self-tuning is very complex in nature and sensitivity analysis is very important for different parameters, since high sensitivity parameters need high degree of tuning.

3.3 Bayesian Case Based Method

Yeguas et al. in [24] proposed a Bayesian case based reasoning method for parameter tuning. The methods discussed in previous sections have the problem of time requirement to evaluate parameters iteratively to get optimize parameter values and interaction of these parameters. This paper evaluates the performance of parameter tuning system empirically to avoid these problems. They combine the Bayesian Networks and Case-Based Reasoning (CBR) method to set optimize parameter values hence maximize the performance of computational intelligence algorithms. In [25, 26] a Bayesian CBR system was introduced as a generic method for solving the parameter tuning problem. The main characteristics which make CBR a good solution for the tuning problem are:

- It is preferable when there is no more information available about the behavioral parameters. This approach uses the past experiences for setting the parameter values.
- It is suitable in problems for which a completely accurate solution does not exist.
- It is suitable when problem instances are very similar in most of the cases.
- The Bayesian CBR system does not require complete information. Furthermore, as the system increases its knowledge, the results improve.

This approach has two important properties, its learning capability, to adapt itself to changes and its capability for autonomy. The design of the Bayesian CBR system consists of mainly two phases. The first phase uses the Bayesian Networks (BNs) relationships among the different parameters. The second phase integrates BNs within a CBR system to solve new problem instances using important features and learning from past for similar problem instances.

3.4 Bi-level Optimization Approach

In early seventies Bracken and McGill [27] introduced bilevel optimization approach in mathematical programming domain. After that numerous studies have been done on bilevel optimization [28, 29]. Most of these bilevel optimization algorithms used nested approach. In nested approach there are two level of optimization, lower level optimization solve the optimization problem for higher level optimization. At lower level Karush-Kuhn-Tucker (KKT) conditions used to transform bilevel optimization into a single level constrained based optimization problem.

In bilevel optimization, there are two nested levels of optimization tasks. Outer one is known as upper level optimization task and inner one is known as lower optimization task. This algorithm has a constraint that lower level optimal solutions can only acceptable as possible input to the upper level. There are mainly two kinds of variable, lower level and upper level. Those are used at lower level optimization task and upper level optimization task respectively. Some characteristics of bilevel parameter optimization are as follows:

- Two nested levels of optimization task.
- Computationally efficient and work well when no. of parameters are high.
- The bilevel optimization converges fast towards optimal parameter settings.

3.5 Additive Procedures

For parameter tuning methods there are some extra additive procedures which increase the capability and search efficiency of parameter tuning methods. These additive procedures are independent of main tuning methods and work as a supplement to main tuner. These additive procedures are given as follows:

- **Racing**: Racing was introduced by Maron and Moore [30]. The racing is used to decrease the number of test cases to determine the quality of parameter vectors, and thus decrease the total runtime of tuning algorithm. The main idea behind racing is that, the number of test cases to determine the utility of a parameter vector is not constant throughout the search. Initially only few test cases are used for each vector and separate out those vector which perform good and increase the number of test cases for those vectors which are not performing worse or better than the good vectors. This approach reduced the significant number of test cases as used in each vector perform tests for each test.
- **Sharpening**: It was introduced by Bartz-Beielstein et al. [31] in Sequential Parameter Optimization (SPO) method. The aim of this sharpening method is also to reduce the number of test cases used to determine the quality of parameter vectors as compared to simple test approach. Initially tuning algorithm start with little number of test cases per vector, as it reaches a certain threshold value, the number of test cases per vector increased to double. Therefore the algorithm explores the search space very fast. This means at the time of termination, the current vector is tested frequently. This leads to very promising results.

4 Algorithm Quality: Performance and Robustness

In computational Intelligence optimization algorithms the accuracy of optimization algorithm is highly dependent on the parameter setting. The performance is highly affected by the optimal parameters setting but parameter optimization is computational overhead and time consuming task. There are several measure to check the performance of computational intelligence techniques some of them are discussed as below.

4.1 Performance Measures

Generally we can check the performance of CI techniques by their solution quality (accuracy) and computational time complexity (speed). The most common performance metrics used in nature inspired computing are as follows:

- Mean of best fitness values over evaluations.
- Average no of evaluations to be performed to get optimal solution.
- Success ratio.

These measures are not always appropriate. If we have a problem which spread of data is very high (large variance), then the algorithm's performance results are questionable. In these cases, mean (and standard deviation) have no significant meaning and it is advisable to use median instead of mean or the best fitness value [32].

Obviously, the actual performance metrics determines the choice of the best parameter settings. Recent studies showed that, different performance measures affect the optimal parameter settings [33]. Without considering the different performance metrics we can't claim anything about optimal parameter settings.

4.2 Robustness

Here robustness means how the performance of optimization algorithm varies over different input parameters. However the performance of any metaheuristic algorithm depends on: problem instance, the parameter vector and the random seed of stochastic process. So there are basically three type of robustness according to problem instance, the parameters and random seed.

If the parameter tuning for an CI algorithm is done using one function over some parameter vector and for some instance of problem do well, it is not necessary that it work well for other problem instances. For robustness to change in parameter values, it is thoroughly measured per parameter individually. These optimization algorithms are stochastic in nature, since they depend on random generations.

So these experimentations require a number of repeated evaluations with identical setup, but with different random seeds.

5　Conclusion and Discussion

This paper provides a comprehensive study of parameter optimization. Accuracy of any computational intelligence (CI) algorithm is highly dependent on optimal settings of parameters. We have discussed how parameter setting affects the performance and robustness of evolutionary algorithms. There is a big question that if we are comparing different parameter optimization algorithm over some benchmark algorithm then whether benchmark algorithm tuned too. Not tuning benchmark algorithm itself is very unreasonable and is biased in nature. So it is equally important to tune benchmark algorithm also tuned before comparison. This paper concludes that parameter optimization depends on various factors and best solution for some problem instance may or may not give optimal solution for other problem instance. Robustness and performance are equally important as optimal parameter settings. In literature there are not much automated tools for parameter optimization so it is open area to develop automated tools for parameter optimization is today's need.

References

1. Y.J. Zheng, S.Y. Chen, Y. Lin, W.L. Wang, Bio-inspired optimization of sustainable energy systems: a review. Math. Probl. Eng. (2013)
2. X.S. Yang, *Engineering Optimization: An Introduction with Metaheuristic Applications* (Wiley, 2010)
3. A.H. Gandomi, X.S. Yang, A.H. Alavi, Cuckoo search algorithm: a metaheuristic approach to solve structural optimization problems. Eng. Comput. **29**(1), 17–35 (2013)
4. D.H. Wolpert, W.G. Macready, No free lunch theorems for optimization. IEEE Trans. Evol. Comput. **1**(1), 67–82 (1997)
5. A.E. Eiben, Z. Michalewicz, M. Schoenauer, J.E. Smith, Parameter control in evolutionary algorithms. IEEE Trans. Evol. Comput. **3**, 124–141 (1999)
6. A.E. Eiben, J.E. Smith, *Introduction to Evolutionary Computation*. Natural Computing Series (Springer, 2003)
7. J. Maturana, F. Lardeux, F. Saubion, Autonomous operator management for evolutionary algorithms. J. Heuristics **16**, 881–909 (2010)
8. A.E. Eiben, S.K. Smit, Parameter tuning for configuring and analyzing evolutionary algorithms. Swarm Evol. Comput. **1**, 19–31 (2011)
9. F. Lobo, C. Lima, Z. Michalewicz, *Parameter Setting in Evolutionary Algorithms*. Studies in Computational Intelligence, vol. 54 (Springer, Heidelberg, 2007)
10. O. Kramer, Evolutionary self-adaptation: a survey of operators and strategy parameters. Evol. Intell. **3**, 51–65 (2010)

11. O.W. Samuel, G.M. Asogbon, A.K. Sangaiah, P. Fang, G. Li, An integrated decision support system based on ANN and Fuzzy_AHP for heart failure risk prediction. Expert Syst. Appl. **68**, 163–172 (2017). Elsevier Publishers
12. A.K. Sangaiah, A.K. Thangavelu, X.Z. Gao, N. Anbazhagan, M.S. Durai, An ANFIS approach for evaluation of team-level service climate in GSD projects using Taguchi-genetic learning algorithm. Appl. Soft Comput. **30**, 628–635 (2015)
13. A.K. Sangaiah, A.K. Thangavelu, An adaptive neuro-fuzzy approach to evaluation of team-level service climate in GSD projects. Neural Comput. Appl. **23**(8) (2013). doi:10.1007/s00521-013-1521-9. Springer Publishers
14. A. Fialho, Adaptive operator selection for optimization. Ph.D. Thesis, Université Paris-Sud XI, Orsay, France (2010)
15. A.K. Qin, V.L. Huang, P.N. Suganthan, Differential evolution algorithm with strategy adaptation for global numerical optimization. Trans. Evol. Comput. **13**, 398–417 (2009)
16. R. Mallipeddi, P. Suganthan, Differential Evolution Algorithm With ensemble of Parameters and Mutation and Crossover Strategies, in *Swarm, Evolutionary, and Memetic Computing*. Lecture Notes in Computer Science, vol. 6466 (Springer, Berlin, 2010), pp. 71–78
17. Z.H. Zhan, J. Zhang, Adaptive Particle Swarm Optimization, in *Ant Colony Optimization and Swarm Intelligence*. Lecture Notes in Computer Science, vol. 5217 (Springer, Berlin, 2008), pp. 227–234
18. R.E. Mercer, J.R. Sampson, Adaptive search using a reproductive metaplan. Kybernetes **7**(3), 215–228 (1978)
19. J. Grefenstette, Optimization of control parameters for genetic algorithms. IEEE Trans. Syst. Man Cybern. **16**(1), 122–128 (1986)
20. T. Bäck, *Parallel Optimization of Evolutionary Algorithms*. Lecture Notes in Computer Science, vol. 866 (Springer, Berlin, 1994), pp. 418–427
21. V. Nannen, A. Eiben, A method for parameter calibration and relevance estimation in evolutionary algorithms, in *Genetic and Evolutionary Computation Conference* (2006), pp. 183–190
22. E.M.H. Pedersen, Tuning & Simplifying Heuristical Optimization, PhD thesis, University of Southampton (2010)
23. X.S. Yang, S. Deb, M. Loomes, M. Karamanoglu, A framework for self-tuning optimization algorithms. Neural Comput. Appl. **23**(7–8), 2051–2057 (2013)
24. E. Yeguas, M.V. Luzón, R. Pavónc, R. Lazac, G. Arroyob, F. Díazda, Automatic parameter tuning for evolutionary algorithms using a Bayesian case-based reasoning system. Appl. Soft Comput. **18**, 185–195 (2014)
25. E. Yeguas, R. Joan-Arinyo, M.V. Luzón, Modeling the performance of evolutionary algorithms on the root identification problem: a case study with PBIL and CHC algorithms. Evol. Comput. **19**, 107–135 (2011)
26. R. Joan-Arinyo, M.V. Luzón, E. Yeguas, Parameter tuning of PBIL and CHC evolutionary algorithms applied to solve the root identification problem. Appl. Soft Comput. **11**, 754–767 (2011)
27. J. Bracken, J. McGill, Mathematical programs with optimization problems in the constraints. Oper. Res. **21**, 37–44 (1973)
28. B. Colson, P. Marcotte, G. Savard, An overview of bilevel optimization. Ann. Oper. Res. **153**, 235–256 (2007)
29. S. Dempe, J. Dutta, S. Lohse, Optimality conditions for bilevel programming problems. Optimization **55**(5–6), 505–524 (2006)
30. O. Maron, A. Moore, The racing algorithm: model selection for lazy learners. Artif. Intell. Rev. **11**, 193–225 (1997)
31. T. Bartz-Beielstein, K.E. Parsopoulos, M.N. Vrahatis, Analysis of particle swarm optimization using computational statistics, in *Proceedings of the International Conference of Numerical Analysis and Applied Mathematics (ICNAAM 2004)* (2004), pp. 34–37

32. T. Bartz-Beielstein New experimentalism applied to evolutionary computation. Ph.D. Thesis, Universität Dortmund (2005)
33. D. Goldberg, *Genetic Algorithms in Search, Optimization and Machine Learning* (Addison-Wesley Longman Publishing Co., Inc., Boston, MA, USA, 1989)

The Maximum Power Point Tracking Using Fuzzy Logic Algorithm for DC Motor Based Conveyor System

Chitra Venugopal and Prabhakar Rontala Subramaniam

Abstract Photovoltaic (PV) generation is largely recognized around the world since it is a renewable resource and it offers advantages such as no fuel costs, less maintenance, zero noise and pollution emission. The specific control and maximum operation of the output power of the solar module is thus crucial and it is required to control the maximum power point tracking for the solar array in a PV module. The objective of this paper is to design a conveyor belt system driven by DC motor whose speed is controlled by solar powered converter operating at Maximum Power Point. In this research, a comparative study of widely-adopted MPPT algorithms such as perturb and observe, incremental conductance algorithm are compared with fuzzy logic algorithm. The results indicated that fuzzy logic algorithms has more stable output power at maximum power operating condition. The PWM signal generated by this method is implemented as a gating signal to boost converter and DC motor drive circuit switches. In the real time implementation, a 50 W rated solar panel is used to draw power and supplied to DC motor drive circuit via boost controller to drive the conveyor system. The designed system is tested using Matlab. After adjusting parameters for real time implementation, the tested circuit is implemented practically to drive the conveyor system with added load at certain intervals. The results shows that the maximum power is extracted from the solar panel at all available times to drive the boost converter and dc motor drive circuit. The measurement of voltage and current at different loading and irradiance conditions shows the practical implantation of the proposed algorithm is successful.

C. Venugopal (✉)
Discipline of Electrical Engineering, University of KwaZulu-Natal, Durban 4004,
South Africa
e-mail: devchith@yahoo.co.uk

P.R. Subramaniam
Discipline of Information Technology, University of KwaZulu-Natal,
Pietermaritzburg 3201, South Africa
e-mail: prabhakarr@ukzn.ac.za

© Springer International Publishing AG 2017   115
A.K. Sangaiah et al. (eds.), *Intelligent Decision Support Systems
for Sustainable Computing*, Studies in Computational Intelligence 705,
DOI 10.1007/978-3-319-53153-3_7

Keywords Maximum power point tracking · Perturb and observe · Incremental conductance · Constant voltage · Fuzzy logic control · Boost converter · Pulse width modulation · DC motor

1 Introduction

As each year passes by, it becomes more evident that the world needs to switch from fossil fuelled supplies of energy to more sustainable, renewable energy sources. One of the major considerations for this type of source is using solar power. Solar power has many advantages; it is clean, sustainable and will never run out. These pros, however, have not totally convinced major power users to focus more efforts on renewable sources of power because of the relatively cheap availability of fossil fuels. One major drawback of solar power is the extremely high initial cost. Presently, however, only about 10% of the world's electrical energy comes from renewable sources.

PV or solar panels are made from the semiconductor silicon, which is the second most abundant element available in the earth's crust. The solar panel is made of multiple connections of PV cells. In a basic, subatomic analysis, PV cells produce electrical energy when light enters the PV cell using what is known as the photovoltaic effect.

The simplest model of a PV cell comprises an ideal current source together with an ideal diode. The current source signifies the current generated by the photons (*Iph*). Since this model is seen as ideal, the output is constant and seen to operate under both constant temperature and irradiance. PV cells are generally characterized by two key factors, the short circuit current and the open circuit voltage. More complex models may be determined by adding resistances in series and in parallel to the system. Solar cells display non-linear current-voltage (I-V) characteristics. Modelling of a solar module is key to understanding the behavior of the module under variable environmental conditions. The efficiency of the module depends greatly on the irradiance level and the ambient temperature at any given time.

Maximum Power Point Tracking (MPPT) can be defined as the automatic control algorithm to adjust the power interfaces in order to achieve the best possible power output from a solar panel. Solar panels exhibit inconsistent levels of power outputs, constrained by unavoidable natural phenomena such as irradiance of light, shading, temperature variations and are also dependent on the quality of the PV module and its own characteristics.

The voltage versus current and power characteristics curves are used to identify a specific operating point at which maximum possible power is delivered [1]. The maximum power point tracking methods can extract more 97% of the PV power when it is properly used. Since the location of Maximum Power Point (MPP) is constantly changing with temperature and irradiance, the MPP needs to be located by a possible efficient tracking algorithm. In this research four maximum power point tracking algorithms are studied, simulated using MATLAB tool box and their results are compared and analyzed carefully [2]. These algorithms include perturb and observe, incremental conductance, fuzzy logic controller and the constant voltage method [1].

The MPPT controllers generally use a switched mode power supply, such as a buck, boost or buck/boost converter. The converter topology required is dictated by the solar panel parameters, battery size and load type. MPPT techniques often employ a controller programmed with the MPPT algorithm. The controller takes measurements such as panel voltage, current, temperature and/or irradiance levels at real, instantaneous time and uses these readings to perform the algorithm. The controller output is the Pulse Width Modulation (PWM) switching signal required to drive boost converter and DC motor drive circuits.

In DC motor drive the MPPT algorithm plays a major role in determining the switching signals for boost converter and DC motor drive circuit. The MPPT technique used was the fuzzy logic controller and it was able to track the maximum operating point fast and accurately. One major drawback of P&O algorithm is that it cannot be fully stabilized at the MPP [3–5]. On the other hand the IC algorithm complicated to implement when compared to the Perturb and Observe technique [6]. Also the measurement of the PV cell's voltage and current can be difficult [7]. In order to improve the performance of boost converter by delivering accurate PWM signal to the switches and to harvest the maximum energy from the solar panel, fuzzy logic method is proposed as the MPPT algorithm. The fuzzy based MPPT method is implemented and tested in DC motor conveyor system. The MPPT tracking using fuzzy logic controller is proved to be accurate and precise because this system has optimum power transferred to the motor and reduced the total harmonic distortion.

The main objective of this research is to drive a conveyor belt system controlled by solar powered DC motor which is driven by MPPT controller. The focus of the research is to analyze the commonly used MPPT algorithms and compare it with fuzzy logic based MPPT controller. In this paper, the P&O and IC MPPT algorithms are discussed with the simulated results. The method of implementing fuzzy logic based MPPT algorithm is shown. This research also focusses onto implementing fuzzy logic based MPPT controller to generate switching signal to control the speed of the DC motor in conveyor belt system. The PWM signal generated by the fuzzy based MPPT system is used to drive the converter switches. The boost converter is governed by the MPPT algorithm to provide a constant 24 V output to drive the motor and battery. The DC motor is to be attached to the conveyor belt which is designed to carry loads of up to 1 kg. The results of MPPT algorithms are compared using voltage, current and power ratings of each method. The design and results of conveyor system and speed control of DC motor is discussed in the following sections.

2   Maximum Power Point Tracking System

The maximum power point tracking (MPPT) is a technique used in wind turbines and photovoltaic (PV) solar systems to maximize power output. Every PV panel has its own I-V pattern because of manufacturing tolerance, shading difference, dust deposited angular displacement in mounted position [6]. This results in the different working operation and has the different maximum power points. A typical solar panel has its own optimal operation point called maximum power points (MPP) which is greatly very as good as cell temperature and sunlight [8]. Various methods and techniques have been implemented to overcome this problem of getting maximum power for optimum efficiency [7]. The complexity relationship between the temperature and total resistance in solar cells produces a non-sequential output efficiency which usually analyzed based on the I-V curve.

$$P_{max} = \frac{V_{max}}{I_{max}} = R_{MPP} \tag{1}$$

where, V_{max} and I_{max} is the PV max voltage and current respectively.

A generic I-V characteristic of PV module shows the maximum power point (MPP) is as shown in equation below [9].

The point of maximum power point is shown in Fig. 1. The slope of the graph is used to track MPP. Since V_{max} and I_{max} vary in accordance with climate changes particularly temperature and irradiance of sunlight. It is always critical to set a maximum power point. To get a maximum power point, a technique is required which can locate the exact position of MPP called MPPT algorithm. The solar panel and MPPT simulations are done using MATLAB. The simulation is performed under the linearly increasing irradiance varying from 200 to 1000 W/m2 in the modelled Tenesol TO 505 panel.

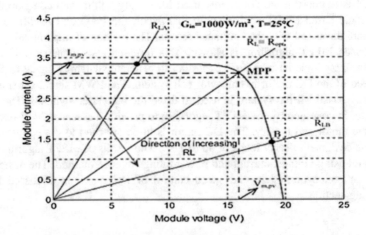

Fig. 1 I-V curve characteristics of a PV cell [9]

2.1 Perturb and Observe (P&O) Algorithm

The P&O algorithm is widely used due to its simplicity and easy implementation. Power is the product of voltage and current. In this algorithm, the operating voltage of a PV cell is little modified/perturbed by small increments and the resulting change in the power (ΔP) is observed [7]. The voltage perturbation is moving toward the maximum power point if the ΔP is positive. This means that the further voltage perturbation of the PV cell will drive the operation towards the maximum power point. If the difference in powers is negative, this means the operating point has moved away from the maximum power point, to return back to maximum power point the direction of perturbation should be reversed [8–10]. The perturbation of voltage in P&O algorithm is shown in Fig. 2.

When the MPP is approached as shown in Fig. 2, the PV voltage is oscillating toward the optimal value of maximum power point. If the step size is large, the MPPT algorithm react quickly to the environmental changes/conditions but if the step size is small, the MPPT is relatively slow and results in not able to react quickly to the change of temperature and irradiance [6, 8].

This algorithm has the four cases, these case are relative to change or perturbation of the voltage while observing the change to power [6].

- When $\Delta P < 0$ & $V(k) > V(k-1)$, then $Vref = V(k + 1) = V (k) - \Delta V$
- When $\Delta P < 0$ & $V(k) < V(k-1)$, then $Vref = V(k + 1) = V (k) + \Delta V$
- When $\Delta P > 0$ & $V(k) < V(k-1)$, then $Vref = V (k + 1) = V (k) - \Delta V$
- When $\Delta P > 0$ & $V(k) < V(k-1)$, then $Vref = V (k + 1) = V (k) + \Delta V$

These four cases as shown in the flowchart in Fig. 3.

The P&O algorithm flow chart shown in Fig. 3 calculates the output power *Ppv* (k) by using the output voltage *Vpv* (k) and output current *Ipv* (k) from the output terminal of solar panel. Then the calculated output power *Ppv* (k) is being compared with the previous step of power calculation *Ppv* ($k - 1$) in order to locate the maximum power point [7]. The difference of powers is zero once the maximum power point is reached [9]. If the different value is not zero, then the next step of calculation will search position of power point at the right hand side (RHS) or left hand side (LHS) of the P-V curve by considering the different value between

Fig. 2 P-V curve of P&O algorithm [3]

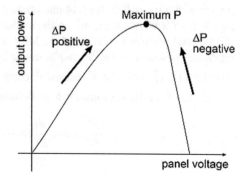

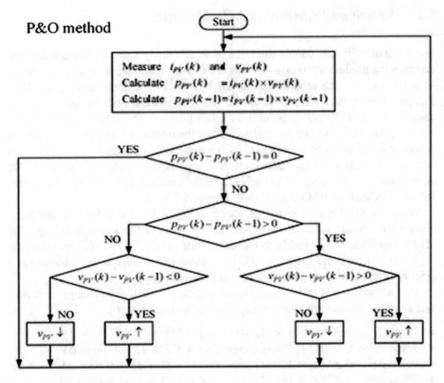

Fig. 3 P&O algorithm flow chart [7, 9]

$Vpv\ (k)$ and $Vpv\ (k-I)$ [7]. The P-V curve using P&O algorithm is shown in Fig. 4.

2.2 Incremental Conductance (IC) Algorithm

This method is based on the fact that the output power derivative of PV cell with respect to the PV cell voltage, at maximum power point is zero [7]. By referring to Fig. 2, the PV panel characteristic shows that this derivative is negative at the right of MPP and positive to the left of the MPP. This lead to conclusive method that the maximum power point (MPP) can be tracked using the comparison of instantaneous conductance $\left(\frac{I_{pv}}{V_{pv}}\right)$ to the incremental conductance $\left(\frac{dI_{pv}}{dV_{pv}}\right)$ [3]. The mathematical modelling of the IC algorithm can be defined by the Eqs. (2) to (4).

$$\frac{dP}{dV} = 0 \rightarrow \frac{dI}{dV} = -\frac{1}{V} \quad at\ MPP \tag{2}$$

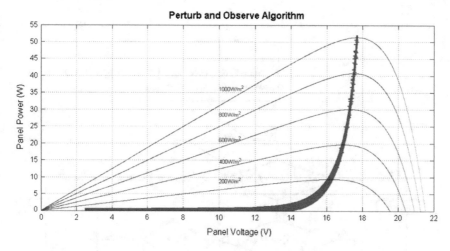

Fig. 4 P-V characteristic of perturb and observe algorithm on Tenesol TO 505 panel

$$\frac{dP}{dV} > 0 \rightarrow \frac{dI}{dV} > -\frac{1}{V} \quad \textit{left of MPP} \tag{3}$$

$$\frac{dP}{dV} < 0 \rightarrow \frac{dI}{dV} < -\frac{1}{V} \quad \textit{right of MPP} \tag{4}$$

Using this mathematical model, the controller will continually measure the parameters of the solar panel and check whether the resultant derivative equates to zero. If not, the controller will decide in which direction a perturbation must occur in order to bring it to the MPP. In this way, the system will be able to clearly identify where the MPP lies at all times and maintain its position until required to change.

The simulated results of IC algorithm is shown in Fig. 4.

The simulations results shown in Figs. 4 and 5 show the application of the P&O and IC algorithms on the solar panel. The algorithms show the tracking of the maximum power point for the range of irradiance levels. It can be seen that both methods are capable of tracking the MPP fairly accurately. The MPP for extremely low irradiance levels usually lies within a range of voltages, and thus there is less convergence at these points. As the irradiance levels increase, the algorithms effectively track the MPP. The P&O method displays much more oscillation around the MPP when compared to the IC. This is because the P&O never truly stays at MPP, rather it oscillates around it within a small range, as is shown. The IC method makes provision to stay at a certain level once MPP is found, thus it shows a better tracking plot. But, IC algorithm is a complicated technique to implement when compared to the Perturb and Observe technique [10]. Also, the performance of this algorithm is poor when dealing with the low levels of irradiance as shown in Fig. 6.

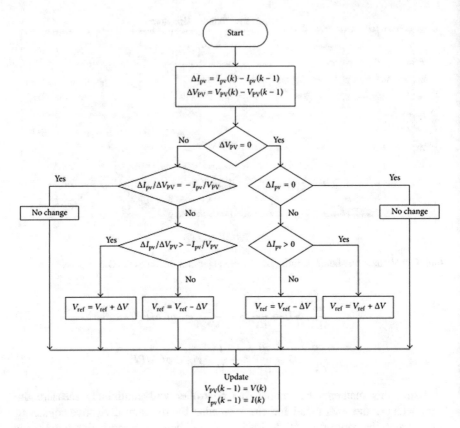

Fig. 5 IC algorithm flow chart

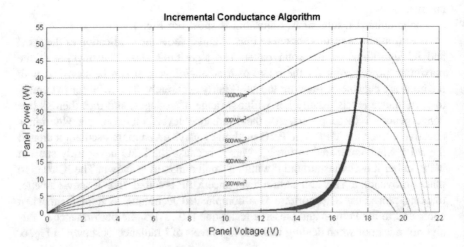

Fig. 6 P-V characteristic of IC algorithm

3   Fuzzy Logic Based MPPT Algorithm

Fuzzy logic is used widely in applications ranging from consumer products to industrial process control to automotive applications [11]. This is mainly because fuzzy logic reasoning is closer to human thinking and it uses natural language rules than conventional logic systems [11–15]. In MPPT, the maximum power point tracking algorithm is also expected to respond accurately to the changes in the ambient conditions that impact the PV output. As shown in Fig. 3, the commonly used P&O algorithm suffers from the drawback of oscillating around the MPP thus creating noise, on the other hand, the IC algorithm tracks the MPP closer to the knee point of I-V curve. In order to overcome the inaccuracies in P&O and IC algorithms, fuzzy reasoning method is implemented to harvest maximum power from the solar panel.

The fuzzy logic system is based on the theory of fuzzy sets and uses linguistic variables. The inputs are transformed into fuzzy sets using membership functions and this process is called fuzzification. The fuzzified inputs are then processed by a set of fuzzy rules to produce the output and this is called Fuzzy Inference System (FIS). There are two FIS systems available namely: 1. Mamdani FIS and 2. Sugeno FIS [16]. At the output level, Mamdani system uses distributed fuzzy set and Sugeno system uses constant or linear output membership functions [17].

The inputs to the fuzzy controller are error and change error values. The controller observes the pattern of these two input signals and generates corresponding output signal using fuzzy inference system. The fuzzified inputs and output are usually handled in per unit (pu) form by using respective scale factors. The output signal is then integrated to generate the actual control signal.

3.1   Design of Fuzzy Logic Controller

The design of fuzzy controller involves design of membership functions for input and output variables. The fuzzy rules are designed based on the PV output power in terms of its voltage and current values. To track the maximum power point, Fuzzy Controller (FC) with voltage and current inputs is used. The fuzzy input variables are current error and voltage error. The output variables of the voltage and current values. The input variables are divided into five membership functions such as negative maximum (NN), negative minimum (NM), zero (ZE), positive minimum (PM) and positive maximum (PP). The output voltage variable is divided into three membership functions such as voltage increase (VP), no change in voltage (VZ) and voltage decrease (VN). Similarly, the output current variable is divided into three membership functions such as current increase (IP), no change in current (IZ) and current decrease (IN). The algorithm continuously updates the voltage and current values to stay at the knee point of the PV curve always.

The input and output membership functions of FC is shown in Figs. 7, 8, 9 and 10.

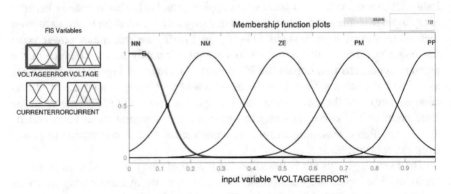

Fig. 7 Membership function of input variable voltage error

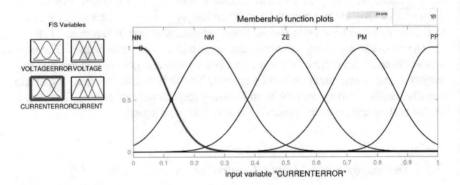

Fig. 8 Membership function of input variable current error

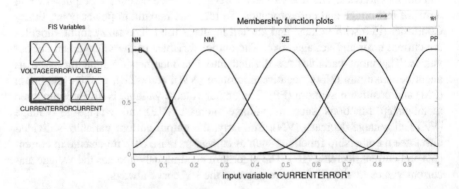

Fig. 9 Membership function of output variable voltage

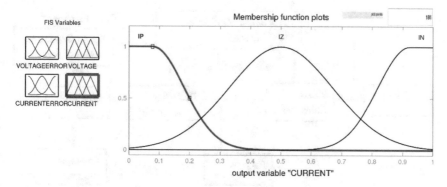

Fig. 10 Membership function of output variable current

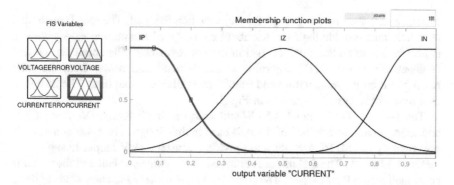

Fig. 11 Surface view of rules

The surface view of the rules is shown in Fig. 11.

4 Design of DC Motor Drive Circuits and Conveyor System

The design is implemented in real time for the prototype model of a conveyor belt system driven by DC motor. The load to the conveyor belt is altered at every time and speed control of DC motor is achieved. The control circuit used to control the speed of the motor is solar powered with chosen MPPT algorithm. The general overview of the design shows the use of a solar panel with maximum power point tracking (MPPT). The boost converter and the DC motor drive circuit are operated based on the PWM signal generated based on the fuzzy based MPPT algorithm. The DC motor drives a conveyor belt load. The speed feedback is obtained and compared with the reference speed to maintain the speed of the conveyor belt

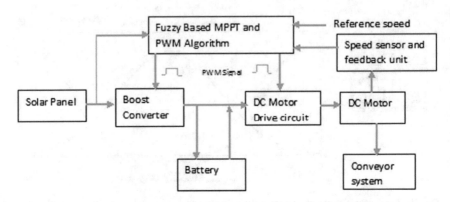

Fig. 12 General overview of the system

system within the operating range for any variation the load. The speed sensor and feedback units provide the feedback about motor speed and current values when it is altered due to add/removal of load in the conveyor belt. The speed and current feedback drives the MPPT algorithm to derive maximum power from the solar panel based on the temperature and irradiance available at that time. The complete overview of the system is shown in Fig. 12.

The Tenesol TO 505, rated at 50 W with an open circuit voltage (Voc) of 21.7 V and short circuit current (Isc) of 3.1 A is used in this design. The boost converter is governed by the MPPT algorithm to provide a constant 24 V output to supply the motor and battery. The DC motor is attached to the conveyor belt and the speed is controlled by the PWM signal. The conveyor belt is designed to carry loads of up to 1 kg.

4.1 Design of Boost Converter

The boost converter system is designed using the solar panel output parameters. The selection of the inductor and capacitor sizes of the boost converter is critical to achieve the desired output from the converter. A switching frequency of 20 kHz is chosen, along with an output ripple voltage and current of 4% and 20% of the outputs respectively. The inductor and capacitor may be calculated using the Eqs. 5 and 6.

$$L = \frac{D\left(1-D\right)^2 V_o}{2f_s I_o} \tag{5}$$

$$C = \frac{DI_0}{\Delta V_o f_s} \tag{6}$$

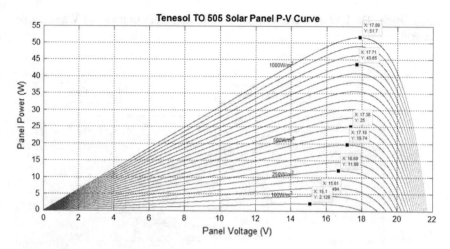

Fig. 13 P-V curve for the entire irradiance range

Since the source power is a solar panel, the input current and voltage are variable and continuously changing. Using just the open circuit voltage and short circuit would not suffice and thus the inductor and capacitor need to be chosen in order to handle the entire range of V_{mpp} and I_{mpp}. The entire irradiance curve of the modelled solar panel is shown in Fig. 13.

Using the data from the modelled solar panel tabulated, the inductor and capacitor may be calculated. The largest value for each is used to take worst case scenarios into account when designing. The values were calculated to be: L = 0.98 mH and C = 56 μF.

4.2 DC Motor Drive Circuit

To perform the mechanical work, a suitable DC motor needed to be selected. The motor chosen is the 37JB6 K, rated at 24 V, 0.08 A and 2300 rpm. The motor is built with a 100:1 gear ratio and thus is able to provide higher torques at slower speed with the same voltage and current rating. The speed after the gear ratio is rated at 23 rpm and has a torque of 0.16 Nm (1.63 kg.cm). The motor torque shows that it is capable of driving the conveyor belt load since it is greater than 0.0927 Nm. The motor is specifically chosen for its high torque and low current. The speed of the DC motor is controlled via a PWM signal from the microcontroller. The duty cycle is to be in the range of 0 to 100% in order to provide speed control of 0 rpm to rated rpm. For the control of the motor via PWM, the L298N is the preferred option which is used in the design. The L298N is a dual, full bridge integrated circuit (IC) driver which is designed to accept standard TTL logic levels and drive motor loads. The L298N driver is capable of driving loads of up to 46 V and 2A per channel.

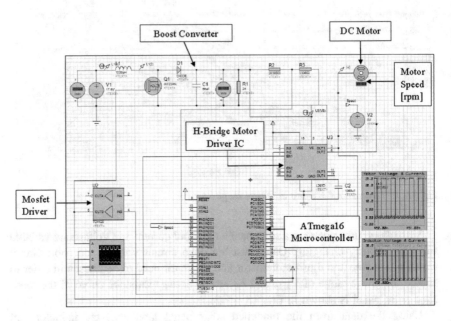

Fig. 14 Proteus simulation model of boost converter, DC motor and drive circuits

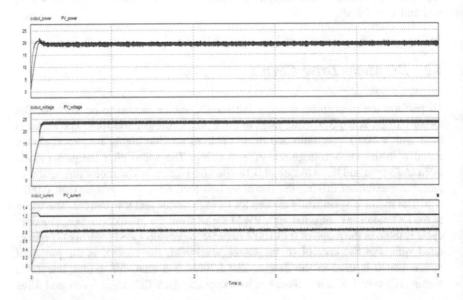

Fig. 15 P&O MPPT curve

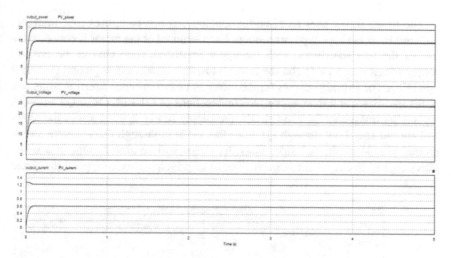

Fig. 16 IC MPPT curve

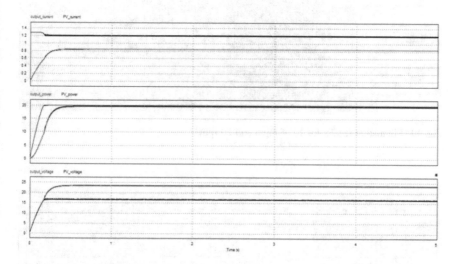

Fig. 17 Fuzzy logic based MPPT curve

Table 1 MPPT comparison

Method	Power (in W)		Voltage (in V)		Current (in A)	
	Panel output	Controller output	Panel output	Controller output	Panel output	Controller output
IC	20	15	16	24	12	0.6
P&O	20	20 ± 0.5	16	24	12	0.8 ± 0.3
Fuzzy	20	20	16.5	23.5	12	0.81

   The motor driver uses an H-bridge topology, which comprises four transistor switches in an H-shape type connection. Each terminal of the motor is connected to two of the four transistors, with the upper transistors connected to the positive supply (drain if npn FET) and the lower transistors connected to ground (source if npn FET). The H-bridge allows for the motor to be run in either direction using two of the transistors at a time; either the combinations of Q1 with Q4 or Q2 with Q3. By turning transistors Q1 and Q4 ON, the supply can be applied across the motor and current may flow through Q1 from the 'A' side through the armature and to ground via Q4 of the 'B' side. Thus the motor will be able to spin in a single direction, say clockwise for example. The same principle may be applied to Q2 and

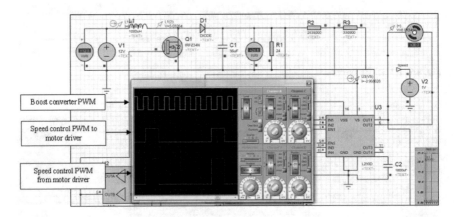

Fig. 18 Simulation results for 12 V input and 20 rpm

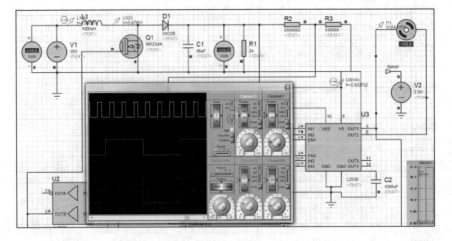

Fig. 19 Simulation results for 15 V with 50 rpm

Q3, spinning the motor in the opposite direction. In essence, the opposite diagonal transistors have to be in the same state whilst the switches on the same vertical line should not be ON at the same time. If Q1 and Q2 (or Q3 and Q4) are ON at the same time at any point, a short circuit will be created. In real world application, it is impossible to switch between current paths via the transistors without momentary equivalent ON states on the vertical switches, thus the need for a diode across each switch to prevent a short circuit of the supply [18]. The conveyor belt system required only one direction of rotation, thus there was no need to switch between the transistor pairs. The simulation of boost converter with DC motor and drive circuit is shown in Fig. 14.

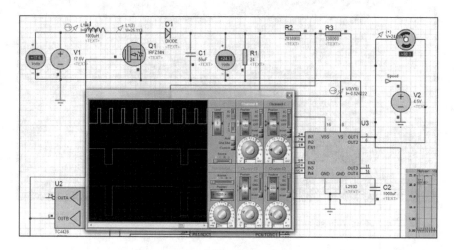

Fig. 20 Simulation results for 17.6 V input with 90 rpm

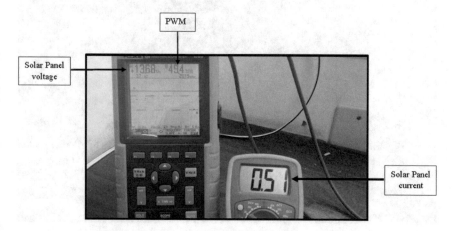

Fig. 21 Solar panel voltage and current under cloudy condition

5   Results and Analysis

5.1   Comparative Analysis of MPPT Algorithms

The implemented controller results for P&O, IC and fuzzy algorithm are shown in
Figs. 15, 16 and 17. These results are obtained using the same DC-DC boost

Fig. 22 Solar panel voltage and current under partial cloudy condition

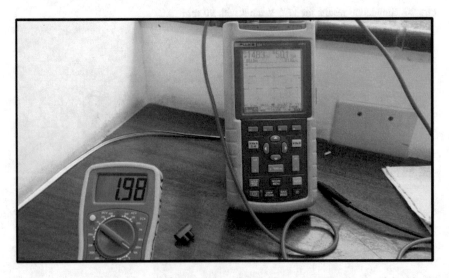

Fig. 23 Solar panel voltage and current under sunny condition

converter circuit operated at the same conditions, constant temperature of 25 °C and constant irradiance of 1000 W/m2.

It can be seen from the results that the IC method gives an insufficient power. The maximum power obtained by using IC method is 15 W and it far away from meeting the desired specifications. P&O algorithm almost performs like an IC method only that the output power is not exactly at the MPP and it has oscillation due to continuous perturbations. The maximum power that can be extracted from the PV solar panel used is 20 W. The comparison of the results for the three MPPT methods used in this project are shown in Table 1.

From Table 1, it can be seen that IC method results are very poor. In P&O algorithm, it can be seen that the algorithm never reaches MPP. The fuzzy method performs results are accurate in terms of output voltage, current and power of the controller. Thus this method is used to drive the DC motor and conveyor belt system.

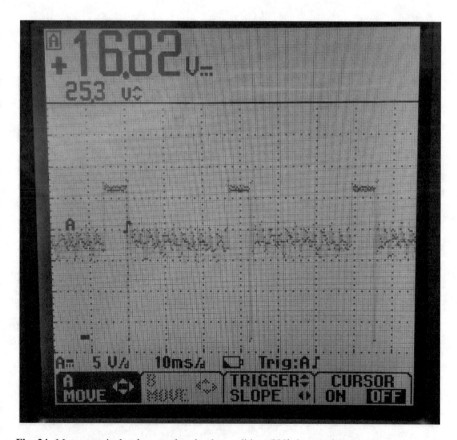

Fig. 24 Motor terminal voltage under cloudy condition (20% duty cycle)

5.2   DC Motor Conveyor and Converter System

The converter circuit simulated using Proteus simulation software is shown in Fig. 14. The operation of the circuit under different speeds and boost converter inputs are shown in Figs. 18, 19 and 20. The simulation with a 12 V input and running the motor at 20 rpm is shown in Fig. 18. The 15 V input with 50 rpm set is shown in Fig. 19, while Fig. 20 represents a 17.6 V input with 90 rpm motor speed set. On the oscilloscope results, the yellow plot indicates the duty cycle provided for the boost converter, the blue line represents the duty cycle for speed control from the microcontroller to the L293D motor driver while the pink graph shows the output of the driver to the DC motor. As can be seen, the PWM for the boost converter progressively regulates itself to maintain a constant 24 V output throughout. These simulations results show that the fuzzy code and components chosen are suitable for the practical application.

The real time measured solar panel voltage and current at cloudy, partially shading and sunny conditions are shown in Figs. 21, 22 and 23.

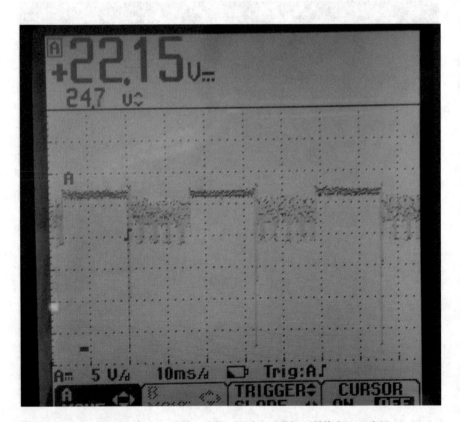

Fig. 25 Motor terminal voltage under partially cloudy condition (50% duty cycle)

The system is implemented in real time for the prototype design of conveyor belt driven by DC motor. The motor voltage variations under these condition at different duty cycles are shown in Figs. 24, 25 and 26. The results show non-perfect PWM waveforms at the motor terminals which can be explained due to a number of reasons. At the positive switching of the PWM, or the ON time, the voltage is approximately 24 V as expected. At the instance that the PWM turns OFF, the voltage drops down to 0 V momentarily but then picks up almost instantaneously. In Fig. 24, this OFF time voltage is approximately 12 V while it is approximately 20 V in Fig. 25. This is due to the regenerating nature of the motor and back EMF. As per the theory, when the motor is rapidly pulsed it creates an on/off effect of the motor. If the motor was constantly supplied with 24 V, it would rise to its rated speed, however because of PWM, the motor picks up speed during the ON time but starts to brake during the OFF time. Since DC machines can be used either as motors or generators, the motor starts to generate a voltage during the braking times at the OFF cycles because of the regenerative braking effect.

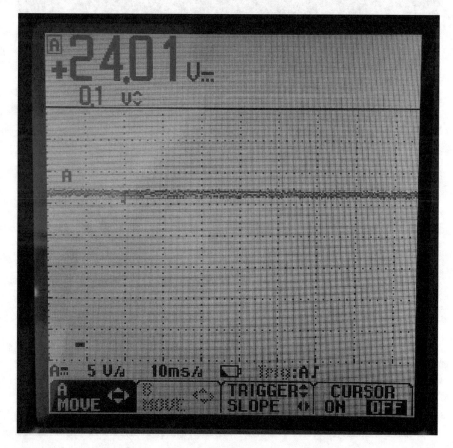

Fig. 26 Motor terminal voltage under sunny condition (100% duty cycle)

The motor current was measured under variations of no load, 0.5 kg load and 1 kg on the conveyor belt system is shown in Fig. 27. The motor initially starts off with the conveyor belt without a load on the belt. The current measured during this time is 0.1 A. At 15 s, a load of 0.5 kg is added to the conveyor belt. The load is applied and rolled off the conveyor belt three times until approximately 33 s.

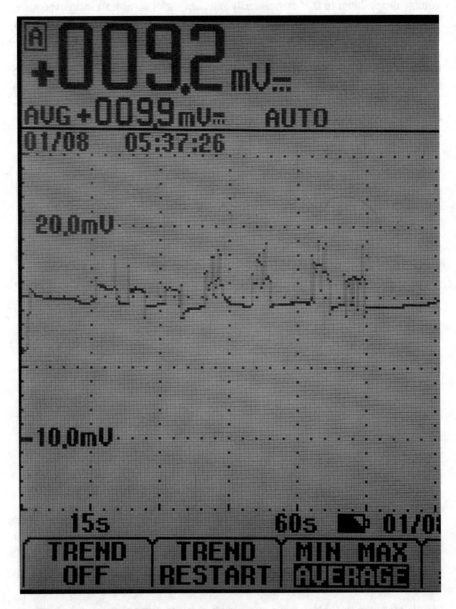

Fig. 27 Motor current at different load conditions

During these loadings the current rises to 0.12 A with spikes of 0.18 A. The conveyor belt is then loaded with a 1 kg load four times, starting at 40 s. It is noted here that the motor current rises and stays at a higher level for longer than the 0.5 kg load. The multiples of current spikes seen during the loading times are due to the imperfect tensioning of the conveyor belt. The belt has slight sag on the bottom layer and thus there is a dip when the load is applied. Because of this dip, the motor has to apply more force to move the load over each of the four rollers, which correspond to the four spikes on each of the 1 kg loading measurements. The average current which the motor draws from the system to pull the conveyor is approximately 0.1 A.

6 Conclusion

The conveyor belt system driven by DC motor is simulated, designed and tested in real time system. The speed control of DC motor is achieved by solar powered converter. The switching algorithm of the converter is controlled by the MPPT algorithm to deliver maximum power, voltage and current to the DC motor based on the temperature and irradiance of the solar panel and load torque variation. In this research, the MPPT algorithms for P&O and IC methods are implemented and compared with fuzzy logic method. It can be seen from the results that the P&O algorithm oscillates around the maximum power point. Hence the results are not accurate. On the other hand, the IC exhibits poor performance at low irradiance level. The fuzzy logic method proves to be accurate, less complicated and less noise level at the output. It can be observed from the results that the fuzzy logic method provides more stable power output at the maximum power operating conditions. The boost converter and DC motor drive circuits are designed to drive the conveyor system. The PWM signal generated by the MPPT algorithm drives the converter circuits. The design is tested using Proteus simulation software and then implemented in real time system. The load analysis is performed by varying the torque added or removed to the conveyor system. The measurement of voltage and current values under different loading conditions shows the successful implementation of design.

Acknowledgements I wish to thank Mr. Sipho Sithole, Mr. Zakhele Ngubane and Mr. Rischav Pillay their contributions in designing and testing of the system.

References

1. G.R. Walker, Evaluating MPPT converter topologies using a MATLAB PV model. *AUPEC 2000: Innovation for Secure Power*, vol. 1 (2000), pp. 138–143)
2. P. Singh, D. Palwalia, A. Gupta, P. Kumar, Comparison of photovoltaic array maximum power point tracking techniques. Int. Adv. Res. J. Sci. Eng. Technol. 2 (2015)

3. M.A. Eltawil, Z. Zhao, MPPT techniques for photovoltaic applications. Renew. Sustain. Energy Rev. **25**, 793–813 (2013)
4. Z.M. AbduAllah, O.T. Mahmood, A.M. T.I. AL-Naib, Photovoltaic battery charging system based on PIC16F877A microcontroller. Int. J. Eng. Technol. **3**(4), 27–31 (2014)
5. A. Harish, M.V.D. Prasad, Microcontroller based photovoltaic MPPT charge controller. Int. J. Eng. Trends Technol. **4**(4), 1018–1021 (2013)
6. A.P. Agung, S. Huda, A. Wijaya, Speed control of DC motor with PWM method using IR control based on ATmega16 mircocontroller, in *International Conference on Smart Green Technology in Electrical and Information Systems* (2014), pp. 108–112
7. M.A. El-Sayed, S. Leeb, Evaluation of maximum power point tracking algorithms for photovoltaic electricity generation in Kuwait, in *International Conference on Renewable Energies and Power Quality (ICREPQ'14)*, no. 12 (2014), pp 1–12
8. S.E. Babaa, M. Armstrong, V. Pickert, Overview of maximum power point tracking control methods for PV systems. J. Power Energy Eng. **2**, 59–72 (2014)
9. P. Suwannatrai, P. Liutanakul, P. Wipasuramonton, Maximum power point tracking by incremental conductance method for photovoltaic systems with phase shifted full-bridge dc-dc converter, in *International Conference on Electrical Engineering/Electronics Computer Telecommunications and Information Technology (ECTI-CON)* (2011), pp. 637–640
10. D.K. Chy, M. Khaliluzzaman, Measuring efficiency of buck-boost converter using with and without modified perturb and observe (P&O) MPPT algorithm of photo-voltaic (PV) arrays, in *International Conference on Mechanical Engineering and Renewable Energy* (2015)
11. A.M. Atallah, A.Y. Abdelaziz, R.S. Jumaah, Implementation of perturb and observe MPPT of PV system with direct control method using buck and boost converter. Int. J. Emerg. Trends Electr. Electron. Instrum. Eng. **1**(1), 31–44 (2014)
12. A. Kaur, A. Kaur, Comparison of Mamdani-type and Sugeno-type fuzzy inference systems for air conditioning system. Int. J. Soft Comput. Eng. **2**(2) (2012)
13. O.W. Samuel, G.M. Asogbon, A.K. Sangaiah, P. Fang, G. Li, An integrated decision support system based on ANN and Fuzzy_AHP for heart failure risk prediction. Expert Syst. Appl. **68**, 163–172 (2017) (Elsevier Publishers)
14. A.K. Sangaiah, A.K. Thangavelu, X.Z. Gao, N. Anbazhagan, M.S. Durai, An ANFIS approach for evaluation of team-level service climate in GSD projects using Taguchi-genetic learning algorithm. Appl. Soft. Comput. **30**, 628–635 (2015)
15. A.K. Sangaiah, A.K. Thangavelu, An adaptive neuro-fuzzy approach to evaluation of team-level service climate in GSD projects. Neural Comput. Appl. Springer Publishers, **23**(8) (2013). doi:10.1007/s00521-013-1521-9
16. M.A.W. Salman, N.I. Seno, A comparison of Mamdani and Sugeno inference systems for a satellite image classification. Anbar J. Eng. Sci. 296–306
17. E.H. Mamdani, S. Assilian, An experiment in linguistic synthesis with a fuzzy logic controller. Int. J. Man Mach. Stud. **7**(1), 1–13 (1975)
18. T. Takagi, M. Sugeno, Fuzzy identification of systems and its applications to modeling and control. IEEE Trans. Syst. Man Cybern. **15**, 116–132 (1985)

Differential Evolution Based Significant Data Region Identification on Large Storage Drives

Nitesh K. Bharadwaj and Upasna Singh

Abstract In today's scenario, almost every user involuntarily generates and utilizes several Gigabytes and Terabytes of data. It is due to the accessibility of diverse and inexpensive digital hard disk drives (HDDs) that have facilitated users with comparably large storage capacities. Almost every digital crime is directly or indirectly associated with storage devices. The ever increasing storage strength of HDD has elevated the forensic examination cost and complexities for the digital forensic investigator. The considerable amount of time is consumed during identification and analysis phase of Digital Forensic (DF) process which creates huge backlog of cases, as a result remarkable delay occurs for availing justice from judicial body. In this research, we propose a methodology to identify forensically significant data regions of suspected drive that can be helpful in accelerating overall digital investigation process. A proof-of-concept technique is developed that utilizes Differential Evolution (DE) for determining the significant data regions and data storage pattern of HDD. The proposed approach incorporates DE which internally utilizes the geometry information of the HDD, i.e. cylinder, track and sector values, for population generation and decision making. Throughout the paper DE samples are defined using the geometry information and entropy as fitness value. Storage devices with different storage capabilities were considered for the experiment and analysis. Detailed case study using the analysis on formatted suspected storage drives highlights the relevance of the proposed approach. The end result is series of output files, providing information about significant regions of the HDD, using which investigator can easily interpret and analyze the suspected drive. Finally, the proposed method is compared with the important functionalities of existing approaches.

Keywords Computational intelligence · Evolutionary computation · Differential evolution · Digital forensics · Storage drive

N.K. Bharadwaj (✉) · U. Singh
Department of Computer Science and Engineering, Defence Institute
of Advanced Technology, Pune 411025, India
e-mail: nitesh_pcse14@diat.ac.in; niteshb2k14@gmail.com

U. Singh
e-mail: upasnasingh@diat.ac.in; upasna.diat@gmail.com

© Springer International Publishing AG 2017   139
A.K. Sangaiah et al. (eds.), *Intelligent Decision Support Systems
for Sustainable Computing*, Studies in Computational Intelligence 705,
DOI 10.1007/978-3-319-53153-3_8

1 Introduction

Gordon Moore in 1965 visualized that of transistor count on an integrated circuit doubles approximately every two year, which has brought revolutionary development in speed, area and capacity of modern processors as well as storage-devices. As a result the technology has become cheaper and grown at exponential rate with reference to Moores law, which has finally converged everything towards digital world. Due to relentless scaling in device size and cost the use of digital devices, for example smart-phone, tablet, laptop, personal computers, camera etc., has been completely dissolved in our daily lives. Necessarily, every digital device cannot be utilized until equipped with memory devices, for example random access memory (RAM), secure digital card (SD), micro SD card, solid state drive (SSD), flash memory card, hard disk drive (HDD), universal serial bus (USB) etc. These storage devices act as a prerequisite to facilitate trending technological benefits to both personal and commercial users.

Every digital device has distinct architecture, configuration and functional capabilities, but the core element of device that enables user and system specific operation are storage devices. Along with the advanced facilities provided by modern digital devices, sometimes it also proves to be a major concern from the perspective of cyber-crimes and unethical activities of offenders. A cyber-crime is defined as a criminal activities carried out by means of computer or digital devices or the internet. The digital devices have created new criminal arena for cyberwarfare, cyber terrorism, fraud and financial crimes, cyberextortion etc. Concurrently, the technology also provides revolutionary platform to criminals and offensive group (cyber and non-cyber) to multiply their influence, for example unauthorized access (hacking), child pornography, electronic harassment, extortion, drug trafficking etc. Possibly every user activities data or information are stored in configured storage devices. Proliferation of digital devices in every domain has created new opportunities and investigative challenges for digital investigators across the world [1]. On the other hand, the examination of suspicious or criminal activities on the seized devices is carried out by digital forensic investigator. The responsibility of investigator is to prove or disapprove the existence of reported suspicious activity based on the forensically examined digital evidence and artifacts. The digital evidence is information (data having forensic value) found on wide range of electronic devices that is equivalent to digital fingerprint of the suspected system. However, the primary objective of digital forensic investigator is to convince the judicial body for justice by utilizing their available Digital Forensic (DF) technical and management strengths. Digital forensic is the process of collection, identification, preservation, examination, analysis and presentation of digital evidence with respect to reported criminal cases, that are legally and judicially acceptable.

1.1 Motivation and Focus

In the present scenario, where DF practitioners are continuing to face "coming digital forensic crisis" [2] possibly due to the storage device capabilities that are now comparably larger, very complex, can consist of huge amount of unstructured and structured data, can be easily intermixed with variety of devices, operating systems, file systems and, media types. Additionally, every year there is rapid increase in manufacturing of storage devices and advancement in their storage capacity beyond the capabilities of processors and existing forensic tools/software. Since, the investigation time increases with the increase data volume that need to be analysed, investigators have observed exponential growth in number of storage volumes registered for forensic analysis which is matter of concern over past few years [3]. During examination of large storage drive it is infeasible to process every byte of data which in turn consumes exhaustive time of the investigator. In the present era of huge volumes of data which will grow exponentially in future and hence, examination time will correspondingly become out of scope. This chapter focuses on alleviating the processing and examination time of suspected storage drive by utilizing computational intelligence (CI) techniques in order to optimize the overall traditional DF process by exploiting the important region of the drive. The idea is to initially determine the selected significant regions of the suspected drive instead of considering every bytes of suspected drive.

For the first time in this chapter we have utilized drive's structural information with differential evolution algorithm to determine significant regions of the drive for achieving fast forensic examination of storage drive. The objective of this work is to study and analyze the affect of evolutionary algorithm for overall advancement of traditional DF process. The assumption considered in this work is that although the suspected storage drive contains different types of information which are stored according to the availability of free spaces i.e. either fragmented or continuous locations. Among different types of available data the highest entropy data is the target requirement whereas the null data sectors are irrelevant to the investigator. The proposed technique focuses on identification of data sectors which have higher entropy values. Hence, this chapter focuses on fast and efficient evidence location identification by using differential evolution algorithm. The proposed methodology utilizes the differential evolution algorithm for determining and examining only the significant data regions of the disk irrespective of the specific target data. Finally, the investigator is provided with the sector locations of significant data for fast examination of suspected drive contents. The proposed approach can be utilized at digital forensic laboratory with no additional cost. The contributions of this chapter are as followed below:

- Trade-off rectangle is proposed to understand present scenario of DF with respect to the resource-utilization, evidence processing-time, accuracy and today's technological gap.

- Significant data regions identification on storage drive for fast examination with the help of popular evolutionary algorithm. The proposed approach best suits the following scenario:

  - Pattern analysis of the data stored in suspected storage drives
  - Determines sector locations of significant data from storage drives
  - Identifying significant data regions within large storage drives

The rest of the chapter is organized as follows: Sect. 2 provides insight into the digital forensic and its pressing issues along with the introduction to computational intelligence paradigm including the contribution of various evolutionary algorithm towards its wide acceptance in diverse application. Implementation details of DE based proposed methodology is covered in Sect. 3 while experimental setup and analysis results in addition with case study are discussed in Sect. 4. Brief discussion on findings and future scope of the proposed work is provided in Sect. 5. Finally, the chapter is concluded in Sect. 6.

2 Background and Related Work

Digital forensic is a practice of investigation as well as recovery of vital materials found in digital devices often associated with computer crimes. A full forensic examination requires processing of each and every byte of the suspected media to determine what it represents and how it is forensically significant to investigator. Hence, unpredictable time is consumed during full-forensic examination of large storage drives. In literature, for examination of HDD, different researchers and investigators presented their concern and contributed various methodologies, tools and techniques using [2–9]. The survey with respect to published and appreciated research as well as their preferred solution in the field of storage device forensic are provided in [3]. The literature covers the forensic solution in consideration with data mining, increased processing power, distributed processing and, artificial intelligence etc. The authors in [2] presented their concern towards the impact of increasing volume of data and the growing number of devices on DF. The paper [2] presents the method of DF data reduction by selective imaging (DRbSI) where, their proposed methodology presents procedure for collection of only information relevant files and databases. The available predetermined files and database are prime focus of the methodology in [2], but no consideration is made if the relevant files or databases are altered, deleted or formatted. The author in [3, 10] presented that the time required for collection and analysis of full disk imaging process increases with increasing volume of data. In contrast to this digital forensic triage is a recent term which engaged the mind of the researchers. *Digital forensic triage is a partial forensic examination conducted under (significant) time and resource constraints* [4]. Both pros and cons exists for triage, where on one side triage helps to reduces the risk of case backlogs in DF laboratories which in turn reduces the long wait of DF examination results, on the other hand the use of triage tools/software also possess a high risk of evidence being missed during investigation. The possibility when information gathering and analysis is not performed the risk of missed investigative opportunities comes into the picture, which is another drawback of triage. Ayers et al. in [11] discussed that due to increase in volume of data and examination complexity the existing forensic software and tools are becoming inadequate. The available forensic tools and software for examining large storage device are now appealing their scalability issues. In support to this the author in [12] presented the challenges of the next decade, and cited the difficulties in capturing, processing and reporting TBs of data. Moreover, most of the HDD available in market contains traces of previously used data that belongs to the former user. Hence, author in [12] shows fast identification of storage drive contents and to determine whether drive is properly wiped or not. In [13] N. Beebe et al. discusses the scope of data mining techniques, which once utilized in DF as an advanced tool it can provide fruitful results in terms of reduced processing time, less analysis cost, improving quality of information, and pattern extraction. The work in [14] demonstrated the Monte-Carlo filesystem search approach on various storage drives in order to search known files in minimum time. However, the scenario where information of known files are unavailable the search process becomes inefficient. The proposed approach identifies the significant regions of the storage drive for forensic examination irrespective of the target data and file system information.

Furthermore, efforts were made to reduce the evidence processing time with the help of *data reduction* approach. In DF investigation almost every reported case is associated with huge volume of data which is seized for forensic analysis provides great scope towards data reduction technique. The author in [15] stated that for a particular circumstance it is necessary to recognize what information needs to be accumulated such that accurate analysis is achieved. This contradicts the process of investigating everything with the practice of examining only required data which can help investigators to achieve accurate analysis. In this support Jonathan et al. in [6] presented *sifting collectors* approach efficiently images forensically relevant regions from Windows New Technology File System (NTFS) which contain important information about *e.g.* Windows OS files, registry, system metadata, temp files, history, logs, browser artefacts etc. However, the forensic imaging emphasis is more on the system and metadata files of Windows file system whereas large amount of actual data contents are not been considered. The work in [6] presents the acquisition of evidence from windows operating system and supported file system format i.e. NTFS. However, the reliability and feasibility issue in consideration with other operating systems and their supported file system types, for example Unix/Linux with FAT, ext2 or ext4 format, Mac-OS etc., were completely ignored. In this chapter, the proposed methodology follows the footprints of selective imaging and analysis methods by using the differential evolution algorithm. The identified significant regions help the investigator for deciding further course of investigative actions. The proposed approach performs the task irrespective of specified operating system, file system formats and particular files and databases. Moreover, if the suspected drive is deleted or formatted the proposed approach can easily spot the significant sector regions where highest entropy data exists. At the initial phase of investigation if the investigator examine the reported significant region of the suspected drive significant saving in examination time is achieved since the regions having null data will be completely omitted. The exclusion of null data and sectors significantly save the investigator from the analysis of null/zero data. The proposed approach to some extent overlaps with the methods defined in [4, 8]. In the following subsection we provide brief discussion on essential basic building blocks of the proposed approach.

2.1 Digital Forensics

In today scenario digital devices are positively evaluated as an vital evidential proof in judicial body. It is essential for an investigator to understand how digital devices are actively involved in criminal activities as well as the type, format and kind of evidence these devices contains. This requires contents analysis and examination of large storage drive. The conventional DF process takes hours to read entire storage drive while the process become exhaustive when fragmented and deleted files need to be considered. Hence, the extraction and collection of forensically sound digital information from large disk volume is burdensome and time hungry. Therefore, the traditional forensic investigation process cannot fulfil the modern demand due to the perpetually increasing capacity of hard drives. In this support we propose a technological trade-off by comparing accuracy, time complexity, resource utilization, development gap among traditional and ideal forensic process which as illustrated using Fig. 1. The ideal forensic process is primary preference of every forensic inves-

Fig. 1 Proposed trade-off using resource utilization, processing time, technology gap and, accuracy constraints of traditional and ideal DF process

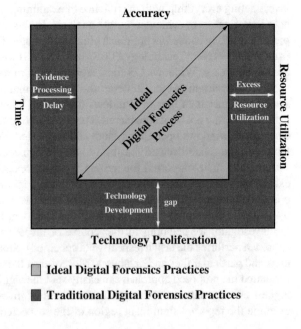

Fig. 2 Hard disk drive basic internal architecture

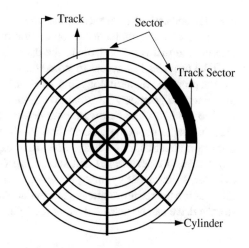

tigator where best forensic analysis result is procured in minimum time with limited resource utilization as represented in Fig. 1. The traditional forensic process proves to be time and resource inefficient that creates huge backlog of filed cases awaiting justice from judicial bodies. However, here we are not contesting to modify existing judicial process. Achieving best situation for forensic investigation requires introduction of new technological achievements and developments.

2.1.1 Storage Disk Drives

The storage disk drive in today's scenario is also termed as hard disk drive (HDD). The data in earlier storage devices like floppy disk is accessible using the geometry and storage structure of the device. The basic geometry of HDD is represented in Fig. 2 using which the data are allowed to write or read in particular sector or set of sectors. The disk comprises a stack of cylinder which spins and allows the drive heads to move over the tracks in order to access particular sectors. The information on a hard disk is stored in tracks, which are concentric circles placed on the surface of cylinder (also known as platter). The head is used to physically locate the location to access the data. The index number of tracks ranges from zero at the edges of cylinder which increases towards center of the track. Sections within each track are called as sectors, which is the basic physical storage unit on a HDD, having 512 bytes size. The cylinder-track-sector (CTS), is an classic method for giving addresses to each physical address of data on a HDD. Though geometrical metrics (platter, track and sectors) no longer have a direct physical relationship to the data stored on modern storage devices except floppy disk, virtual geometrical values (which can be translated by disk driver or software) are being used by many utility programs and file systems to access data by using sector locations. The reason is that less head movement is possible whenever geometry is used for data access. Although, each cylinder consists of two tracks (one on each side), the modern drives utilize only one track of the cylinder while other side of track is used for storing the control information that is unavailable to the operating users. Hence, the geometry metrics of HDD provides fast access to stored data.

The proposed methodology implicitly utilizes *hdparm* command line tool in order to extract the geometry information of storage drives. This command line utility is used to set and view hardware parameters of hard disk drives. Hdparm tool provides valuable geometry information (CTS parameters) about any storage device. For example let us try to derive disk drive capacity which is a function of cylinder (C), tracks (T), sectors (S) and sector size (B). Consider a particular disk has X_1 cylinder, X_2 track per cylinder, X_3 sectors per track and X_4 sectors size in bytes, the information are retrieved after execution of hdparm utility. The total storage capacity storage drive can be provided using the following Eq. 1.

$$HDD\_Capacity = \prod_{i=1}^{4} X_i \ (C, T, S, B) \ bytes \tag{1}$$

In this chapter the CTS information extracted during run time is utilized throughout for identification of significant data regions from suspected storage drives. Therefore, the total capacity of the provided HDD geometry information can be easily derived and also each sector is independently accessed for analysis of data. This chapter utilizes the geometry metrics to access particular data of 512 bytes for searching significant data regions of the disk drive.

2.1.2 Major Issues in Storage Drive Forensics

The easy availability and accessibility of today's large storage drives has provided remarkable features to the criminal users and new challenges to the investigators. The open issues related to examination of storage drives from the perspective of investigator are discussed as followed: Every user now a day involuntarily generates and utilizes several Gigabytes (GB) and Terabytes (TB) of data where it is infeasible for investigator to analyze and process every byte of data within specified time bounds. The continuous development in storage technology i.e. increasing storage capacity in almost all consumer devices, the increasing number of variety of digital devices and cloud storage services has exponentially increased the number of devices seized per case. Digital investigation tools and techniques are lagging behind the pervasive advanced technologies due to which investigator exhaustive time is consumed in the earlier examination phase of traditional DF process. Seizure of number of devices as well as lack of well defined investigation tools creates huge backlogs of evidence awaiting analysis. Time constraint and processing delay also creates huge log of unprocessed cases which in turn develop huge delay in evidence processing as a result corresponding delay in judicial decision is obtained from legal bodies. In this chapter we present our concerns towards above mentioned investigation issues and propose a methodology for evidence locality identification to overall accelerate digital investigation process by utilizing popular evolutionary algorithm. In the next sub-section an insight into the computational intelligence paradigm is provided for better understanding of the proposed approach.

2.2 Computational Intelligence Paradigm

Before proceeding to problem objective it is necessary to understand what computational intelligence really is. Computational Intelligence (CI) is the branch of science and engineering where complex computational problems are solved by modeling problems according to the natural and biological intelligence, resulting in *"intelligent systems"*. These intelligent systems encapsulate numbers of popular intelligent algorithms; artificial neural networks, evolutionary computation, artificial immune systems, fuzzy system and swarm intelligence. Hence, these intelligent algorithms belong to the field of *Artificial Intelligence* (AI). Alternatively, it is stated that computational intelligence is a sub-division of AI, where CI is analysis and study of adaptive mechanism to facilitate intelligent responses in complex and differing environment.

This section highlights the subcategories of computational intelligent paradigm namely evolutionary computation (EC) and differential evolution. Computational intelligence refers to the study and design of intelligent algorithms that has ability to learn specific task and behavior from the experimental data and result observations. CI techniques generally address the complex real-world problems for which conventional mathematical modeling is undefined or unpredictable. The final outcome of computational algorithms acts as a heuristic solution to solve particular objectives related to diverse area of research.

2.2.1 Evolutionary Computation

Evolutionary computation (EC) mimics the nature-inspired evaluation competence, where the major approach of survival of suitably best generation is supported while the weakest is left to die. In natural evolution process survival is achieved with the help of reproduction similarly the offspring is reproduced from two or more parents that contain the best characteristic of each parent. The produced individuals that inherit imperfect characteristics are always weak which lose combat to survive, for example in the bird species out of several one infant manages to get more food, becomes stronger, as a result the strong infant kicks down all weak siblings from the nest to die.

Individual in evolutionary algorithms (EA) is referred to as a chromosome and EA works on large population of individuals with maximum randomization. The characteristics of individuals in the population are defined by the chromosome where each characteristic is called as gene. Each individual of population compete for reproduction of suitable offspring. The offspring with desired values have long survival capabilities and good chances of reproduction for further population generation. Crossover and mutation are the two crucial processes for every evolutionary algorithm. Crossover is the process where offspring is generated by combining the sub-parts of two or more parents while mutation is the process where each offspring undergo alteration of some characteristics of the chromosome. The deciding factor for the survival of produced individual is fitness value that is derived by corresponding objective or fitness function. The constraints of problem and objective are modelled in the form of fitness function. In this way with the use of fitness function, population generation process, crossover process, mutation and selection process, a successful evolutionary computation can be utilized to solve the untouched problem of real world. However, in the literature several EA have been developed to solve several problems as listed below:

- Genetic algorithm: models the chromosome recombination and mutation process for searching heuristic solutions
- Differential evolution: Operates on vector differences hence, best for numerical optimization problems
- Cultural evolution: It is a process of dual inheritance where a set of behavioural traits form population individuals
- Genetic programming: Computer programs forms a set of genes and chromosomes for further reproduction and mutation operations
- Neuroevolution: The evolutionary algorithms are used to train artificial neural networks

In the literature over a period of time the real-world application have been successfully benefited from evolutionary computation, for example, operations research, robotics, combinatorial circuits optimization, process scheduling, fault tolerant systems, chemistry and physics etc. In this chapter the differential evolution algorithm is used to enhance overall DF process especially for examination of large storage drives from digital investigation perspective.

2.2.2 Differential Evolution

A popular evolutionary algorithm (EA) i.e. Differential evolution (DE) was proposed by Rainer Storn and Ken Price in 1997. DE has always provided acceptable results whenever used in a variety of problems from diverse fields [16]. Similar to other EAs, at each generation DE also utilizes mutation, crossover, and selection operations in order to achieve global solution by moving population toward the optimal solution. DE's performance is mainly dependent on two components:

1. Offspring vector generation: mutation and crossover operations
2. Control parameters: population size (NP), crossover control rate (CR), and scaling factor (F).

The details of each component in contrast to the prescribed objective are elaborated in the subsequent section of the chapter. Researchers are continuously utilizing DE for finding optimum solution to crack their research complications. For example, DE has been successfully applied to distinct areas of science and technology, such as signal processing [17], chemical engineering [18], machine intelligence & pattern recognition [19], and mechanical engineering design [20]. The experiments performed over several numerical benchmark problems [21] illustrate that DE performs better than the particle swarm optimization (PSO) [22] and genetic algorithm (GA) [23]. At the initial phase of DE algorithm generates random population by using uniform distribution followed by mutation, crossover and selection operations in order to further generate a new population. The crucial step in DE algorithm is generation of trial vectors. Mutation and crossover operations are the two basic steps for generating the trial vectors. The best trail vector is selected by the selection operator for next population generation. The discussion on implementation details of the proposed methodology based on DE algorithm along with the insight into important sub-operations of DE (crossover, mutation and selection) is provided in the following section.

3 Implementation of Differential Evolution Based Significant Data Region Identification

In this section discussion on DE based proposed significant region identification methodology is provided along with detailed into the internal characteristic of the algorithm. The proposed approach uses the HDD geometry to access every possible random sector data in the manner similar to that is shown in Eq. 2. The proposed technique generates random geometry information (C_{r1}, T_{r2}, S_{r3} values) to get access to random sector location of the HDD.

$$\text{Sector\_Number}_r = C_{r1} \times T_N \times S_N + T_{r2} \times S_N + S_{r3}$$
$$\text{where } C_{r1} \in C_{N1}, \ T_{r2} \in T_{N2}, S_{r3} \in S_{N3}$$
$$f(r1, r2, r3) = rand(N1, N2, N3) \ and,$$
$$N1 = \#\text{Cylinders}, N2 = \#\text{Tracks} \ \& \ N3 = \#\text{Sector}$$

(2)

where, C_{r1}, T_{r2}, S_{r3} and Sector_Number$_r$ are the randomly selected geometry parameters while, C_{N1}, T_{N2} and S_{N3} are the total cylinder, track and sector numbers of the disk drive. Every EA operates with maximum randomization of the parameters for better performance and results therefore, the proposed technique also rely on achieving more random behaviour after selecting random CTS values (C_{r1}, T_{r2} and S_{r3}), as represented in Eq. 2. The CTS values are the important numerical parameters that provide access to every possible data location of storage drive. The fitness function for the proposed approach is *Entropy* (E) of the sector data, the generalized equation for entropy is presented in Eq. 3, where $\forall x \in$ Sector data bytes. The entropy is calculated for every randomly selected sector which generates large amount of sector samples for further processing and analysis.

$$E(x) = - \sum_{\forall x} p(x) \log_2 p(x) \tag{3}$$

The higher value of entropy represents higher forensic relevance of the retrieved sector or the region. The idea is to traverse maximum number of data sectors for deciding the valid data region boundaries. In this direction the CTS numeric parameters of the disk are utilized in generation of the population or trial vectors that is required for DE algorithm. The initial population or trial vector is generated with the selection of random values of CTS and computation of corresponding entropy value for accessed sector, respectively. Here, the initial population size is fixed to 50 hence, it is not guaranteed that all the individuals from population will have entropy greater than 0. The presence of insignificant sectors (null sectors) varies the size of the next generation of population. However, as the number of iteration increases the size of population saturates to the desired number of individuals. In this way each individual and population consists of cylinder, track, sector numbers and corresponding fitness value for analysis of relevant data regions. Moreover, the completion of specified iterations and run of DE algorithm provides retrieved data sector locations having high entropy value. The retrieved data sectors can be manually examined in order to finalize the data relevant regions of the disk drive. The basic DE sub-operations with respect to the storage drive are briefly described as follows:

- Mutation: Each and every individual of the current population generates their corresponding trial vectors with the help of mutation operator. The target vector is mutated with weighted differential for generating the trial vector. Every offspring is the outcome of the crossover operation after using the recently generated trial vector. For example, let us consider N is the generation counter index, the mutation operator for trial vector generation $M_x(N)$ from the parent vector $P_x(N)$ for detailed discussion on the mutation operation as followed below:

  1. The individuals from a population is a function of $M_x(N) \in f(C, T, S)$ and $P_x(N) \in f(C, T, S, E)$
  2. A target vector $P_{x1}(N)$ is selected from the population, such that $x \neq x1$ where, $[x, x1] \in \{C, T, S\}$ one at a time
  3. The two individuals P_{x2} and P_{x3} are also randomly selected from the population such that, $x \neq x1 \neq x2 \neq x3$ where, $[x, x1, x2, x3] \in \{C, T, S\}$ one at a time
  4. The mutation operator proceeds the calculation of next trial vector once the target vector is mutated as described below:

$$M_x(N) = P_{x1}(N) + \overbrace{F \times \underbrace{(P_{x2}(N) - P_{x3}(N))}_{\text{Step size}}}^{\text{Variation component}} \tag{4}$$

where the mutation scale factor is $F \in (0, 1)$ that controls the amplification of the differential variation [24].

- Crossover: Offspring $P'_x(N) \in f(C, T, S, E)$ is generated using the crossover of parent vector, $P_x(N)$ and the trial vector, $M_x(N)$ as follows:

$$P'_{xy}(N) = \begin{cases} M_{xy}(N), & \text{if } y \in Y \\ P_{xy}(N), & \text{Otherwise.} \end{cases} \tag{5}$$

The set of crossover points is represented by Y. Alternatively we can state that, Y is the points that will follow perturbation, $P_{xy}(N)$ which is the yth element of the vector $P_x(N)$.

- Selection: The selection operator consists of two functions as follows:

  1. Selection of the individual for the mutation operation in order to generate the trial vector
  2. Selection of the best among the parent and the offspring based on their corresponding fitness value for the next generation of population.

The fitness value is the deciding factor for the replacement of parent from the population. Whenever the offspring has better fitness as compared to the parent then the parent is replaced from the population otherwise, the parent remains in the population.

$$P_x(N+1) = \begin{cases} P'_x(N), & \text{if } f(P'_x(N) > P_x(N)) \\ P_x(N), & \text{Otherwise.} \end{cases} \tag{6}$$

This ensures that the average fitness of the population does not deteriorate. Algorithm 1 illustrates the pseudo code for DE algorithm based significant data region identification strategy, where scale factor is F, CR is the crossover rate, CTS are randomly selected geometric values of HDD and P is population vector.

Algorithm 1 Pseudo code for the DE based significant data region identification

---
1: Control parameters → number of iteration, CR and F are initialized;
2: Initial population $P(N) \in f(C, T, S)$ is generated;
3: **while** stopping criteria(s) $\neq$ true **do**
4:     **for** each individual $P_x(N) \in$ Population(N) **do**
5:         Evaluate the fitness, $f(P_x(N)) =$ Entropy(sector bytes);
6:         Trial vector is created by using the mutation operator $M_x(N)$;
7:         Offspring $P'_x(N)$ is created by applying the crossover operator;
8:         **if then**$f(P'_x(N))$ is better than $f(P_x(N))$
9:             Add $P'(N)$ to Population($N+1$);
10:             Memorize the individual and related parameters for best fitness value
11:         **else**
12:             Add $P_x(N)$ to Population($N+1$);#*No fitness value is better than previous value*
13:         **end if**
14:     **end for**
15: **end while**
16: Return all the individual with the best fitness value corresponding to each run as the solution;

The proposed approach creates output file corresponding to each run with several iterations of DE algorithm. The investigator can easily interpret the output files and examine only the reported regions of the drive for further analysis. However, it is necessary to take care of total DE evaluation in terms of number of run and iteration such that particular regions of HDD can be significantly analyzed. If the size of HDD is small the decision can be achieved using less number of DE evaluations whereas the large size of HDD requires more number of DE evaluations. A trade-off exists among size of storage drive, number of DE evaluation and accuracy. Along with the number of evaluation other parameters like CR and F also plays vital role in providing feasibility to solve real time problems. Similarly, in this chapter CR and F act as a critical parameter to decide overall efficiency of the analysis to identify data relevant region of the storage drive.

4 Experimental Setup

In order to evaluate the efficacy of the proposed approach several experiments are conducted using DE based significant data region identification approach that aims to overall accelerate evidence processing phase of DF process. The proposed approach is designed and developed under Python 2.7.11 environment installed on Kali Linux 2.0 operating system working over a multi-core desktop. The experimental environments are governed by the following parameter settings of DE based proposed methodology:

- Size of the population NP = 50
- Mutation rate F = 0.5 (default value),
- Cross-over rate CR = 0.5, (default value)
- Maximum number of DE evaluations for which the default value is set to 200000. Similarly, the runtime error or the stopping criterion for the proposed approach are defined in Table 1
- The implicit number of runs = 20 and iteration = 200, and graphs are plotted using the best fitness of each run.
- Data sector read size for fitness computation is 512 bytes

The stopping criteria for proposed approach are tabulated in Table 1, the program execution halts when any of the given criteria is unsatisfied. The analysis for identification of data relevant region on HDD is performed using storage drive of different capacity with different DE parameters as shown in Table 2.

Moreover, different data volumes are stored in the considered storage drives. For example, our experiment utilizes completely filled, completely wiped or partially filled storage drive. Since, every evidence seized under forensic investigation is classified unless directed from judicial body. The demonstration of the proposed approach is presented using a real time synthetic case study as discussed and demonstrated in the later section.

Table 1 Stopping criteria for execution of proposed technique

S. no.	Error	Details
1	Unable to locate the storage drive	The error occurs whenever the storage drive is undetected or the tool is provided with incorrect device ID
2	Unable to traverse output directory	Provided output directory is inaccessible for storing output files
3	Invalid number of runs	The default number of *run* is modified to $run \leq 0$
4	Cross-over rate (CR) is invalid	The default value of CR is altered to $CR \leq 0 \,\&\, CR > 1$
5	Mutation rate (F) is invalid	The default value of F is altered to $F \leq 0 \,\&\, F > 1$

Table 2 Experimental environment results inclusive of DE parameters

Drive size	Parameters	Identified sector regions
4 GB	CR = 0.5, F = 0.5, Runs = 20, Iterations per run = 200	864263, 1420474, 1693669, 1828392, 2320737, 2718215, 3595197, 3844604, 3860357, 4033162, 4148844, 4273448, 4292235, 5514173, 5744091, 6381149, 6782653, 7306662, 7306662, 7435147
8 GB	CR = 0.6, F = 0.4, Runs = 20, Iterations per run = 300	10426, 12661
16 GB	CR = 0.7, F = 0.4, Runs = 25, Iterations per run = 400	42427, 2330292, 3009887, 3885245, 6720428, 6745560, 7060698, 8930829, 9653354, 9653354, 9653354, 10029472, 12401876, 12401876, 12401876, 13660138, 15406446, 16282954, 16436649, 20819511, 20882496, 20882516, 20973998, 21026847, 27336892
1 TB	CR = 0.7, F = 0.3, Runs = 30, Iterations per run = 500	143497698, 284615592, 309868202, 320808730, 323561273, 326772585, 353198522, 429059156, 430042193, 501248142, 568475352, 595183261, 677497286, 759557978, 783639441, 890793828, 909184718, 913227618, 939949440, 941416087, 982055020, 1127464244, 1192365960, 1201495925, 1564215420, 1715947574, 1738762014, 1777650051, 1788991651, 1791885375

4.1   Experimental Results and Analysis

The storage drives with different capacities of 4 GB, 8 GB, 16 GB and 1 TB are used to validate the efficiency of the proposed significant data region identification approach. The proposed technique is evaluated against the considered storage drives using basic sector read size (512 bytes) and parameters as provided in Table 2. The result reveals that the proposed approach successfully reports the sector that consists of data with high entropy values among the processed sectors. Hence, there is high probability that the other sectors located around or close to the reported sectors also contains data having considerable entropy values. The consideration of other nearby sectors with respect to the extracted sectors, form a data region that can have forensic relevant artefacts which can be important for examination and analysis. The data sectors identified after execution of proposed approach are included in Table 2 which have highest entropy (fitness) value corresponding to particular run of DE.

Alternatively, investigator can also traverse the nearby sector locations corresponding to identified sectors for more advanced analysis. However, the drive regions which consist of null/irrelevant data are completely ignored until significant sector are identified. The experiments were performed on storage drives of 4 GB, 8 GB, 16 GB and 1 TB with varying DE parameters as shown in Table 2. Since, 4GB drive is completely filled with random data, more number of relevant sectors have been identified, which consist of almost all range of drive sectors, as illustrated in Fig. 3a. However, second experiment consist of 8GB disk drive which is almost empty or blank therefore, only couple of significant sectors have been identified while other ranges of disk sectors have been ignored due to presence of null data, as shown in Fig. 3b. Similarly, the experiments were performed on 16GB and 1TB disk drive for different parameters the result of which is illustrated in Table 2 and Fig. 4. The proposed technique determines the data sectors which have high entropy values irrespective of the amount of null data present in the suspected drive.

Furthermore, detailed insight into the proposed technique based on the experimental results is provided with the help of Figs. 3 and 4. The x-axis represents sector number while y-axis represents the fitness (entropy) value of corresponding sector data. Analysis on the referenced figures highlights the actual data storage pattern that exists within the suspected disk drive. The data pattern reflects the location and magnitude to data that actually exists within the storage drive. For example, the low entropy data exists at starting and mid of the storage drive whereas the high entropy data are located near the higher magnitude sector locations of the storage drives, as illustrated in Figs. 3 and 4.

The fitness value of initially generated population and final fitness value after completion of each run are also compared to analyze the diversity and convergence behavior of DE evaluation as illustrated in Figs. 5 and 6. The proposed approach identifies the significant data sectors even when a new storage drive is examined, as represented in Table 2 and Figs. 3b and 5b, where most of the fitness values converged to *zero* entropy and very few data relevant sectors have been identified. It might be possible that the reported sectors and its neighbor sector regions contains

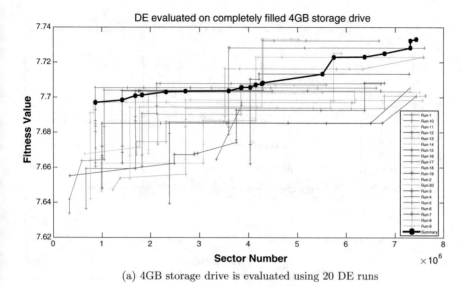

(a) 4GB storage drive is evaluated using 20 DE runs

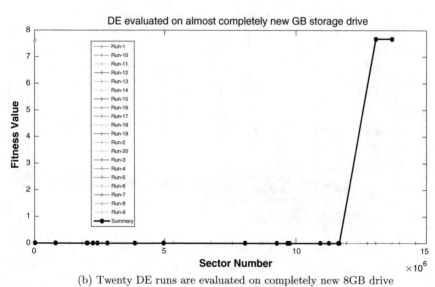

(b) Twenty DE runs are evaluated on completely new 8GB drive

Fig. 3 Fitness values of each iteration is compared with best fitness values for every run of DE evaluation in sorted order. The analysis consists of disk drive with 4 GB and 8 GB storage capacity

metadata or other vital information. Analysis of suspected HDD using the proposed methodology can comparably save examination time of digital investigator. Hence, the proposed approach can help in accelerating overall DF process and investigation by providing insight into the suspected HDDs.

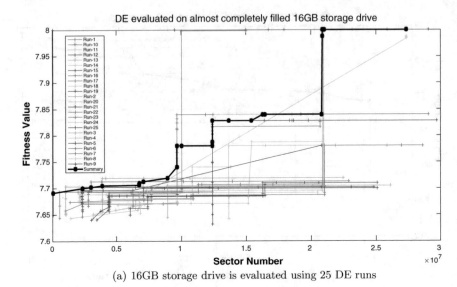

(a) 16GB storage drive is evaluated using 25 DE runs

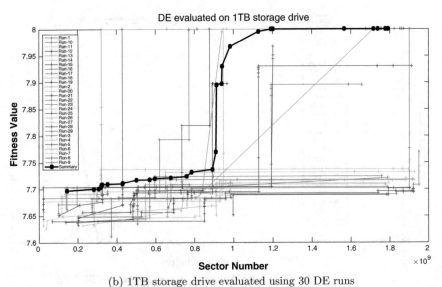

(b) 1TB storage drive evaluated using 30 DE runs

Fig. 4 For 16 GB and 1 TB storage capacity the fitness values of each iteration is compared with best fitness values for every run of DE evaluation in sorted order

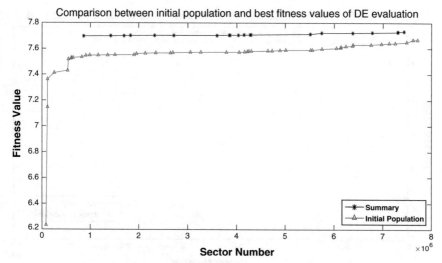

(a) Best fitness of each run compared with initial population
fitness values on 4GB drive

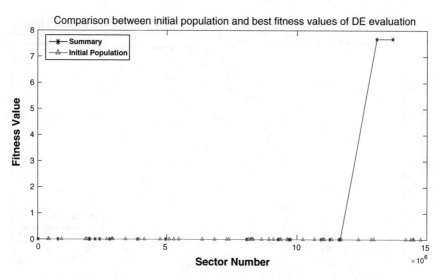

(b) Best fitness of each run compared with initial population
fitness values on 8GB drive

Fig. 5 Comparison between fitness values of initially generated population and fitness values for each run of DE evaluation. The disk drive with different capacity have been analyzed **a** 4GB, **b** 8GB

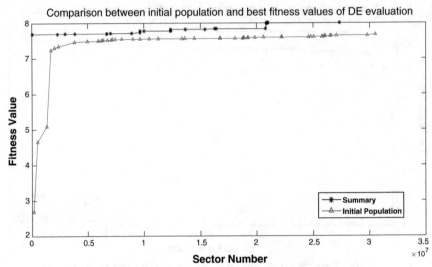

(a) Best fitness of each run compared with initial population
fitness values on 16GB drive

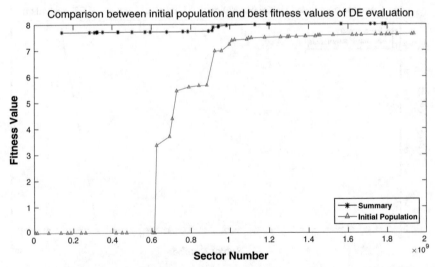

(b) Best fitness of each run compared with initial population
fitness values on 1TB drive

Fig. 6 Comparison between fitness values of initially generated population and fitness values for each run of DE evaluation. The disk drive with different capacity have been analyzed **a** 16 GB and **b** 1TB

4.2 Case Study: Examination of Formatted Storage Drive

Two storage drives having storage capacity of 16 GB and 1 TB are suspected to contain obscene material (child pornography video). Digital investigator has seized the suspected drive for examination. Unfortunately, the seized drives (16 GB and 1 TB) were completely formatted before their seizure. The duty of the investigator is to identify the sensitive materials in order to approve or disapprove the allegation of suspected person.

During investigation the investigator is unaware of the actual contents of the storage drive. Additionally, it is unpredictable that the suspected drive is either properly wiped or completely or partially filled with data. The proposed technique which utilizes the storage drive structural information with other static parameters, such as CR = 0.4, F = 0.6, Runs = 20, NP = 100 and iteration = 300, as an input arguments to DE algorithm for identification of significant data region of the suspected drive. The proposed technique successfully identifies the existence of high entropy data even form the formatted suspected drive. The identified significant regions is provided in Table 3 where, the proposed technique successfully identifies high entropy data regions irrespective of analysis on the formatted drive. As an investigator instead of processing every byte of suspected drive, if identified sector and corresponding regions are examined, significant saving in investigative time and resources can be achieved. The region is extracted with the help of upper and lower bounds of identified sectors which can be represented with $[2^{\lceil log_2(\text{Sector Number})\rceil}$

Table 3 Case study analysis using proposed methodology

Drive size	Identified sectors	Significant regions
16 GB	17988238, 18027376, 18061904, 18074590, 18075466, 18085326, 18198933, 18296372, 18427476, 18478941, 18571921, 18912875, 18945782, 18961698, 19031569, 19036241, 19100409, 19100830, 19270940, 19338096, 19346907, 19362214, 19397889, 19398333, 19398433, 19469614, 19603038	16777217-to-33554431 sector ranges
1 TB	256747015, 346142300, 368445042, 554576085, 818309559, 867492316, 894690345, 900264512, 938450979	134217729-to-268435455, 268435455-to-536870911 and 536870912-to-1073741823 sector ranges

and $2^{\lfloor log_2 (\text{Sector Number})\rfloor}$] respectively, in such a way that no important data can be missed. Hence, the proposed algorithm helps investigator to target the significant regions of the disk during the initial phase of investigation. The examination of identified sector region has revealed the existence of desired data instead of processing the complete disk drive for the investigative analysis. Thus, with the help of DE algorithm significant saving in evidence processing and analysis time can be achieved.

5   Discussion and Future Scope

The continuous advancement in the technology development from past few decades in coordination with increasing volume of data resulted in huge backlogs of digital investigation case and increase in evidence processing time. The significant data region identification method outlined in this work highlights the opportunity to elevate the processing speed of large storage drives forensic examination by utilizing storage drives structural information and differential evolution algorithm. The proposed approach can be applied to the forensic images and physical media to overall reduce the evidence processing time. The comparative summary of functionalities in existing approaches and proposed technique is provided in Table 4. The Table 4 does not compare the methods via experimentation, but variety of approaches with differing methods that were discussed in this chapter is highlighted. The general approaches of various methods were highlighted using the comparison. The future consideration and detailed comparison would be to undertake more insight into the highlighted literature of this chapter.

The proposed significant data region identification method using differential algorithm have demonstrated the potential to speed-up the evidence processing in comparison with full examination of large storage drives. However, the proposed approach is not meant to replace full evidence examination, and there are majority of circumstances where full evidence examination is mandatory. In this paper for the first time computational intelligence paradigm has been utilized to provide vision towards achieving fast digital forensic investigation results. The proposed trade-off rectangle lists out number of problems that need to be focused for advancement of existing digital forensic technology. The major development is required towards the development of novel DF methodology/tools/software in order to mitigate excess resource utilization, evidence acquisition and processing delay and so on. In this chapter we used differential evolution algorithm as a core decision maker, but it is also possible to utilize other evolutionary algorithms that are available under computational intelligence paradigm. The proposed technique outlined in this work locates the forensically significant region of the suspected drive. During investigation process, investigators are not provided with the internal details of suspected disk drives. Hence, the use of proposed technique prior to examination phase can provide data pattern as well as insight into the disk drive. Moreover, performance analysis of the proposed approach with alteration of standard DE parameters can be

Table 4 Comparative summary of existing work and proposed approach functionalities

Functions and methodology	Proposed approach	Full examination	Triage solutions	Garfinkel fast disk analysis	Sifting collector approach	Monte Carlo based search
Significant sectors and region identification	√	√	*	X	X	X
Processing of forensically relevant data from deleted/formatted drive	√	√	–	√	X	X
Classify null and high entropy data region	√	X	X	√	X	X
Processing random sector samples	√	X	*	√	X	√
Target/specific data/evidence processing	X	√	–	√	√	√
Ability to ignore null data and sectors	√	X	*	*	X	X
Human in the loop	*	√	*	–	√	√
Header based search	X	*	*	√	–	–
Reliable to physical media and mounted image	√	√	√	√	*	√
Utilizes structural information of physical media	√	X	X	X	X	X
Not Specific to one file system	√	√	√	√	X	√
Speed-up acquisition, examination and analysis	√	X	√	√	√	√

(√) Addressed issue
(X) not addressed
(*) may be applicable
(−) not necessarily applicable

next step of analysis. The performance and accuracy of the evolutionary algorithm varies with different parameters such as crossover, mutation, number of iterations etc. Therefore, the proposed technique is also provided with the flexibility to customize the execution characteristics and parameters of DE algorithm according to the user need. Hence, the declaration and definition of standard values i.e. CR, F, number of runs and iteration, and population size with respect to storage capacity of drive can be another dimension for future work. The consideration of various storage characteristics such as completely filled drive, completely wiped drive, formatted drive, deleted drives, contiguous file storage and fragmented file storage, can also be utilized for future extension of the proposed methodology. The limitation of the proposed work comes into picture whenever the suspected drive is encrypted or encoded; as a result every sector is identified as significant sector. The encrypted drive can make the proposed method ambiguous. The enhancement of the proposed approach to user interface is also left for future work. The study of the affect of sector read size other than 512 bytes over the proposed methodology is also another dimension of research.

6 Conclusion

In this chapter we presented a new trade-off rectangle to reveal present requirements towards the development of existing DF facilities. The chapter also proposes a methodology that extract data storage pattern using storage drive's structural information and DE algorithm that can help investigator in planning further course of action. Significant saving in investigation time and resource utilization can be achieved by the investigator with the help of proposed technique. The proposed approach also identifies the significant data sectors within storage drives. Although modern HDD provides good storage capability, the proliferation of these devices has also created new hurdles for digital investigators. Presently, the traditional digital forensic process requires considerable time for performing investigative task. This chapter emphasizes on the identification of data relevant sector regions in HDD using computationally intelligent DE algorithm to accelerate the overall DF process. The proposed method enables the examiner with insight of relevant data sector location and data intensity that is present within HDD such that, the investigator can directly access the specified data regions for investigative examination and analysis. Since, high entropy reflect maximum information content the proposed approach provides data locations which have globally high entropy among all the sectors addressed within HDD for particular iteration of DE algorithm. It is preferable to increase the number of iterations with the increase in size of HDD and vice-versa. The proposed approach provides acceptable regions of the suspected HDD, which is under investigative consideration. The further search and analysis can be narrowed down and continued manually once the data relevant regions are identified. Hence, the proposed technique can be of great help to the digital forensic community for fast evidence examination.

References

1. M.G. Williams, A risk assessment on Raspberry Pi using NIST standards. Int. J. Comput. Sci. Netw. Secur. (IJCSNS) **15**(6), 22 (2015)
2. D. Quick, K.K.R. Choo, Big forensic data reduction: digital forensic images and electronic evidence. Springer Cluster Comput. 1–18 (2016)
3. D. Quick, K.K.R. Choo, Impacts of increasing volume of digital forensic data: a survey and future research challenges. Elsevier Digit. Investig. **11**(4), 273–294 (2014)
4. V. Roussev, C. Quates, R. Martell, Real-time digital forensics and triage. Elsevier Digit. Investig. **10**(2), 158–167 (2013)
5. A. Shaw, A. Browne, A practical and robust approach to coping with large volumes of data submitted for digital forensic examination. Elsevier Digit. Investig. **10**(2), 116–128 (2013)
6. J. Grier, G.G. Richard, Rapid forensic imaging of large disks with sifting collectors. Elsevier Digit. Investig. **14**, S34–S44 (2015)
7. S.L. Garfinkel, Carving contiguous and fragmented files with fast object validation. Elsevier Digit. Investig. **4**, 2–12 (2007)
8. S.L. Garfinkel, A. Nelson, Fast Disk Analysis with Random Sampling (2010)
9. N. Kishore, B. Kapoor, Faster file imaging framework for digital forensics. Procedia Comput. Sci. **49**, 74–81 (2015)
10. F. Adelstein, Live forensics: diagnosis your system without killing it first. Commun. ACM **49**(2), 63–66 (2006)
11. D. Ayers, A second generation computer forensic analysis system. Elsevier Digit. Investig. **6**, S34–S42 (2009)
12. S.L. Garfinkel, Digital forensics research: the next 10 years. Elsevier Digit. Investig. **7**, S64–S73 (2010)
13. N. Beebe, J. Clark, Dealing with Terabyte Data Sets in Digital Investigations, in *Advances in Digital Forensics* (Springer, 2005), pp. 3–16
14. J. Dalins, C. Wilson, M. Carman, Monte-carlo filesystem search—a crawl strategy for digital forensics. Elsevier Digit. Investig. **13**, 58–71 (2015)
15. G. Palmer et al., A Roadmap for Digital Forensics Research, in *Forst Digital Forensics Research Workshop*, Utica, New York (2001), pp. 27–30
16. R. Storn, K. Price, Differential evolution—a simple and efficient heuristic for global optimization over continuous spaces. J. Global Optim. **11**(4), 341–359 (1997)
17. S. Das, A. Konar, Two-dimensional IIR filter design with modern search heuristics: a comparative study. Int. J. Comput. Intell. Appl. **6**(03), 329–355 (2006)
18. P.K. Liu, F.S. Wang, Inverse problems of biological systems using multi-objective optimization. J. Chin. Inst. Chem. Eng. **39**(5), 399–406 (2008)
19. T. Rogalsky, S. Kocabiyik, R. Derksen, Differential evolution in aerodynamic optimization. Can. Aeronaut. Space J. **46**(4), 183–190 (2000)
20. M.G. Omran, A.P. Engelbrecht, A. Salman, in *2005 IEEE Congress on Differential Evolution Methods for Unsupervised Image Classification*, vol. 2 (IEEE, 2005), pp. 966–973
21. J. Vesterstrom, R. Thomsen, A comparative study of differential evolution, particle swarm optimization, and evolutionary algorithms on numerical benchmark problems, in *Congress on Evolutionary Computation, 2004. CEC2004*, vol. 2 (2004), pp. 1980–1987. doi:10.1109/CEC.2004.1331139
22. J. Kennedy, R. Eberhart, Particle swarm optimization, in *IEEE International Conference on Neural Networks, 1995. Proceedings*, vol. 4 (1995), pp. 1942–1948. doi:10.1109/ICNN.1995.488968
23. J.H. Holland, *Adaptation in Natural and Artificial Systems: An Introductory Analysis with Applications to Biology, Control, and Artificial Intelligence* (U Michigan Press, 1975)
24. A.P. Engelbrecht, *Computational Intelligence: An Introduction* (Wiley, 2007)

A Fuzzy Based Power Switching Selection for Residential Application to Beat Peak Time Power Demand

Chitra Venugopal, Prabhakar Rontala Subramaniam and Mathew Habyarimana

Abstract The power demand during peak period causes large power fluctuations in the commercial grid. The reduction in using grid supply during peak time benefits the supplier and consumer. There are many methods such as monitoring the usage of grid supply, implementing load shedding, using local generators and disconnecting unwanted loads during peak time, used to reduce grid dependency during peak time. These methods reduce power demand from grid by either supplying interrupted power to the load or by establishing additional power source. The objective of this research is to solve the high power demand from grid during peak time and to avoid the frequent load shedding problem. The research design includes design of hybrid power system. In this research, solar and grid power are chosen as two power sources and automatic switching system to select the power source to supply uninterrupted power to the load is designed. The automatic selection of power source during peak time with selection of power source based on the availability of the source is designed and tested using Proteus real time simulation software. The introduction of fuzzy logic method to select the source based on the utility power level, and availability of the source is also described. The results of the fuzzy logic system in selecting the power source during peak time along with the utilized power level is analyzed. A manual switch is added to the system to override all the input conditions which is also considered as one of the input to the fuzzy decision making system. The designed system is tested in MATLAB and Proteus simulation software. The outputs are displayed in the LCD screen added to the microcontroller. The results displayed are the power source selected based on the

C. Venugopal (✉) · M. Habyarimana
Discipline of Electrical Engineering, University of KwaZulu-Natal,
Durban, South Africa
e-mail: devchith@yahoo.co.uk

M. Habyarimana
e-mail: hmatayo@gmail.com

P.R. Subramaniam
Discipline of Information Technology, University of KwaZulu-Natal,
Pietermaritzburg, South Africa
e-mail: prabhakarr@ukzn.ac.za

© Springer International Publishing AG 2017
A.K. Sangaiah et al. (eds.), *Intelligent Decision Support Systems for Sustainable Computing*, Studies in Computational Intelligence 705,
DOI 10.1007/978-3-319-53153-3_9

availability of the source, peak and off peak time and the power utilized by various sources. The results shows that the designed system is more flexible in selecting the power source based on the utility level and availability of power source to save energy. It can be seen that the designed system eliminates the drawbacks of the existing system to reduce stress on the grid during peak time. Also it is economically feasible to implement it and efficient in supplying uninterrupted power to the load.

Keywords Hybrid renewable energy systems · Load shedding · Hybrid solar and mains · PV panel · MPPT · DC-AC converter · Boost converter · Fuzzy logic · Membership functions · Peak time · Off peak time

1 Introduction

Commercial grid power supply fluctuations largely depend on the peak power demand load. This challenge is further compounded by the fact that most generation plants are located far away from the cities, and the losses that occur during the transmission increase due to the distance factor [1, 2]. The capacity in the supply lines often become insufficient during peak time. The problem is not only raised by the transmission failures during peak times, but also by generation capacity shortages.

In order to help alleviate such fluctuations as well as insufficiency of generated power, suppliers use local power generators in some countries during peak periods but most of these generators are costly oil or gas-fired plants. This has a ripple effect where the price of electricity gets increased considerably for big consumers [3].

The reduction in using grid electricity during peak time benefits both consumers and suppliers. On the consumer side, the price will be reduced and on the supplier side there will be no constraints due to the transmission of electricity during this peak time. As a result, any method that can be used to reduce total dependence on grid power during peak period is highly welcome. Despite many alternative measures taken by the electricity suppliers to reduce the peak time energy demand, it remains a problem due to the reality of the need for electrical power supply during peak time.

This leads us to consider the need for alternative energy sources to be combined with the mains by selecting the appropriate source for utility. Different methods are developed to solve the problem of peak power demand.

The first solution that electricity companies and distribution authorities apply to problems related to peak time power demands is to advise consumers to switch off unnecessary appliances by highlighting the drawbacks of heavy consumption of electricity. These drawbacks include the incremental of the price. In formal public statements and announcements: here "Eskom estimates that HVAC (heating, ventilation, air conditioning) contribute 5400 MW (about 15%) to peak demand." [4].

Nevertheless, this solution seems not favorable due to the following reasons;

a. There are unavoidable consumption loads which can give rise to problems when turned off during peak time. Appliances such as air conditioning and fans constitute the largest consumer of electricity in households [5, 6], especially during winter and or summer seasons. The hotter the day, the longer the fans and/or air conditionings are on and the colder the day the longer the air conditioning needs to heat rooms. Moreover, the installation of air conditioning units in houses increases considerably [7].

b. Additionally, the time for bathing, cooking (recalls Geyser and electric stove) and business activities cannot easily be changed due to the reality of people's need for electricity which climaxes in peak time demand.

The monitoring usage method is an alternative method acquired by electricity suppliers or big consumers, where the monitoring system is applied which predicts the demand for the next period of time. In this system, the alarm is set on at the maximum power demand level. When the set value is reached, an alarm alerts system users to take action. Actions that can be taken are, reducing us- age by means of load shedding or adding the generation capacity [8]. This system reduces the peak power demand when the actions are taken accordingly; ESKOM estimates that understanding power alert makes a measurable difference [9], of course it needs the intervention of the human being and the load is interrupted.

Another important and more frequently practicing method is Load Shedding. The load shedding is to voluntarily stop the supply of one or more consumers to quickly restore the balance between production and consumption network. This is a safeguard measure designed to avoid the risk of voltage or frequency collapse that could cause the cut of an entire subnet.

There are four types of load shedding namely Load shedding by order, Load shedding by energy metering, Load shedding on power threshold and Load shedding on frequency threshold are currently practiced in different countries [10].

In the above listed methods, either the load is disconnected from the source or the alternate power source is used during power failure. These methods could not solve the power demand during peak time. The high cost in addition to the control and supervision of energy storage management system reduces the efficiency of the system.

This research has been conducted to solve the high power demand from grid during peak time and to avoid the frequent load shedding problem.

The Hybrid power system of or simply hybrid system is a combination of different but complementary energy generation systems. Approximately 90% of Hybrid Renewable Energy Systemsn (HRES) studies are on design/economic aspects and have reported it to be highly costly [11]. The cost reduction and technological development of hybrid energy systems in recent years has been encouraging, but they still remain an expensive source of power.

In many applications, research has focused on the performance analysis [12, 13] of demonstration systems and the development of efficient power converters, such as bi-directional inverters, battery management units and maximum power point trackers [14, 15]. Various simulation programs [16] are available, which allow the optimum sizing of hybrid energy systems. However, fewer studies were reported on the design and control of the hybrid solar and mains system.

In this research fuzzy logic is introduced in the selection of automated hybrid solar and mains system during peak time to reduce the stress on the grid. The existing system is designed to select the power source during fixed peak and off peak times. The introduction of fuzzy logic method in this research makes the system flexible within the range of peak and off peak time to save more energy. The design can be extended to select the power source beyond the defined peak times based on the utilized power level. So measurement of power level is introduced in the system.

This objective of this research is to solve the high power demand from grid during peak time and to avoid the frequent load shedding problem. The research design includes design of hybrid power system. The automatic selection of power source during peak time with selection of power source based on the availability of the source is designed and tested using Proteus real time simulation software. The introduction of fuzzy logic method is to select the source based on the peak time, manual switch position, and availability of the source is also described. The results of the fuzzy logic system in selecting the power source during peak time along with the utilized power level is analyzed to implement it practically. The real time parameters indicated in ESKOM's website is used in the control algorithm and design of the controllers. An LCD display is added to the see the results of the controller. The results displayed are the power source selected based on the availability of the source, peak and off peak time and the power utilized by various sources.

2 Automatic Selection of Power Source During Peak Time

The design automatically selects solar panel power from battery to supply 1 kW load during peak power demand period and during mains power shut off period. The selection between solar and mains power is automatic; the system does automatically check the availability of the sources to switch between solar and mains during peak and off peak times.

During night when the mains shuts off, battery is utilized to power the load. The program is user friendly and can be changed any time according to the requirement. Also, for emergency needs, manual switching is provided to change the source. The program is written based on the peak and off peak time electricity demands mentioned in the ESKOM website. The switching algorithm is shown below and the flow chart is shown in Fig. 1. The designed system can work continuously and the monitoring is done on second to second basis depending on the frequency of the used microcontroller.

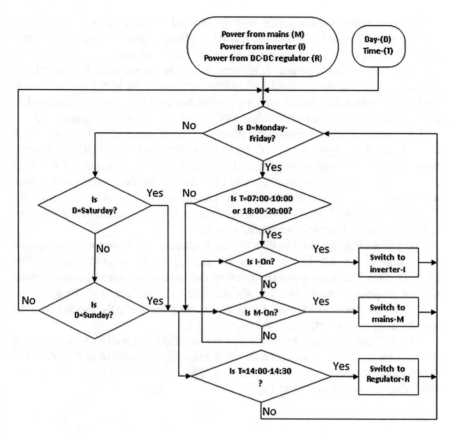

Fig. 1 Power source selection flowchart

Start.

Input 5 variables: Power from Mains (M), Power from inverter (I), Power from DC-DC regulator (R), Day (D), and Time (T)

Step D_1: If Day is Monday to Friday, go to step D_2, else, go to Step D_6

Step D_2: If Time is either 07h00-10h00 or 18h00-20h00 go to step D_3, else go to Step D_4

Step D_3: If Inverter is on, go to A_1, else go to Step D_4

Step D_4: If Power from Mains is on, go to A_2, else go to D_3

Step D_5: If Time is 14:00-14:30, go to A_3

Step D_6:If Day is Saturday, go to D_4, else go to D_7

Step D_7: If Day is Sunday, go to D_4 and D_5, else go to D_1

Step A_1: Switch to the inverter (I), go to Step D_1

Step A_2: Switch to the mains (M), go to Step D_1

Step A_3: Switch to Regulator (R), go to Step D_1

The solar panel receives power from the sun and the solar panel is connected to the Maximum Power Point Tracker (MPPT), a fully electronic system that varies the electrical operating point of the modules so that the modules are able to deliver the maximum available power. Additional power harvested from the modules is then made available as increased battery charge current [17]. MPPT can be used in conjunction with a mechanical tracking system, but the two systems are completely different. In this study the mechanical MPPT system has not been considered.

The power harvested from the PV panel and MPPT is stored in the battery bank and converted through the DC-DC boost converter and through DC-AC inverter then fed to the AC load during the peak time demand. The switching of this solar energy to the load is done automatically by means of the time-based programmed microcontroller switch. Alternatively, the load is supplied by the power from the AC mains of the grid during the Off-Peak time according to ESKOM's data source [18] by means of the same automated switch.

The panel cleaning robot is directly fed from the solar power after being converted through the DC-DC converter for getting the linear output power. The cleaning cycle is also automatically done by the same programmed microcontroller and the program is set according to the panel dirt settlement rate of the area.

Besides the common AC load, the system is designed to supply also the DC load and this supply is directly from the battery bank.

All automatic switches are combined (or installed in parallel) with a manual switches for the user to take action any time manually. In addition to this, the LCD screen displays any activity done instantaneously [19]. This is shown in the general block diagram of the design in Fig. 2.

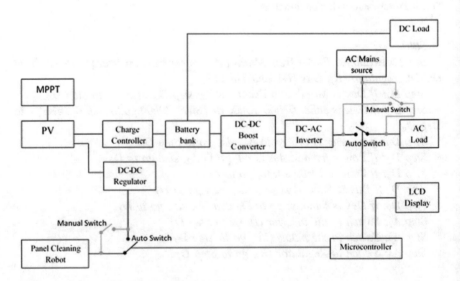

Fig. 2 General block diagram of the design [19]

3 Fuzzy Logic in Automatic Selection of Power Source During Peak Time

Fuzzy logic is used widely in applications ranging from consumer products to industrial process control to automotive applications [20]. This is mainly because fuzzy logic reasoning is closer to human thinking and it uses natural language rules as compared to conventional logic systems. The conventional system selects the power source based on defined peak and off peak times whereas in real time, peak and off peak power consumption may be different for different users on different days of a week. In the power source selection during peak time application, it is required that the system selects the source around the defined peak to save energy. The power saving would be more efficient if the system is designed to select the solar or mains source based on the peak power and also peak time. Hence, fuzzy logic is used to select the source based on the fuzzy rules with peak time, power consumed and the availability of the source. The fuzzy reasoning method occupies less memory in microcontroller. There are two types of fuzzy inference systems such as Mamdani and Sugeno are available. The Mamdani FIS is commonly used due to its simple structure and easily understandable rule base whereas Sugeno FIS is accurate and the output is more adequate for functional analysis. The Mamdani FIS can be used for multi input single output system as well as multi input multi output system whereas Sugeno FIS can only be used in multi input single output system. In this research Mamdani FIS is used because of its capability to defuzzify the fuzzy output using defuzzification techniques.

The input to the fuzzy system are peak times, Solar Power, Mains Power and manual input position. The outputs are solar input and mains input. The input variables "Solar Power", "Mains Power" and "Peak times" are divided into 5 membership functions as NN, NM, ZE, PM and PP as shown in Figs. 3, 4 and 5. The input variable "Manual Switch Position" is divided into three variables namely SOLAR, OFF and MAIN as shown in Fig. 6. The output variable "Power Source Selected" is divided into 7 variables N1, N2, N3, ZE, P3, P2, P1 representing solar

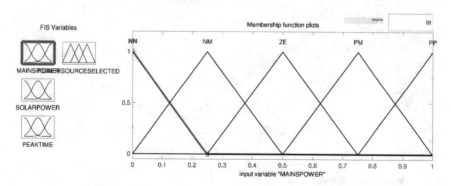

Fig. 3 Mains power fuzzy input

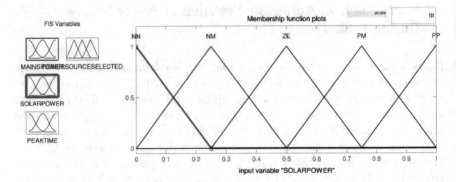

Fig. 4 Solar power fuzzy input

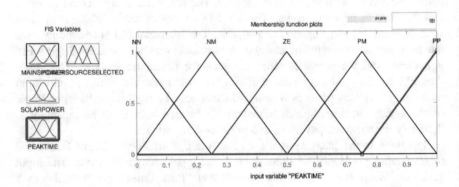

Fig. 5 Peak time fuzzy input

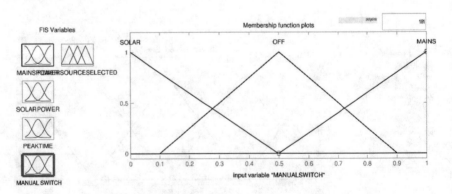

Fig. 6 Manual switch fuzzy input

on, solar off, main on, mains off, solar faulty, mains faulty and complete off as shown in Fig. 7.

The surface view of the fuzzy rules governing the input and output variable to select the appropriate source based on the time and availability of the source is shown in Fig. 8.

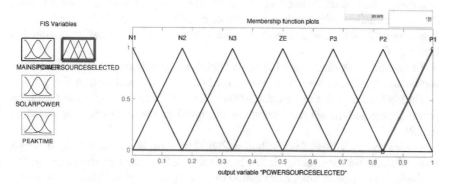

Fig. 7 Power source selected fuzzy output variable

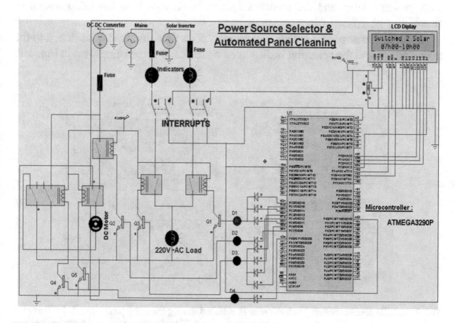

Fig. 8 Simulation model of the system [19]

4   Results and Analysis of the Design

The switching circuit consists of the power source selection between the power
from the mains and solar system. The selection based on the peak time power
demands is done by the programmed microcontroller which also drives the dc
motor for the panel cleaning robot. The circuit is visualised on the LCD screen
indicating every activity done at the time. The designed system is tested using
Proteus simulation software. The peak times considered for the design are 07h00 to
10h00 and 18h00 to 20h00. The Proteus simulation model of the design is shown in
Fig. 8 [19].

During the peak time, from 07h00 to 10h00, the load is switched to solar power
source and the source selection is displayed on the LCD as shown in Fig. 9a.

During the off-peak time, from 10h00 to 18h00 the load is switched to the mains
power source and the source selection is displayed on the LCD screen as shown in
Fig. 9b.

During the peak time, from 18h00 to 20h00 the load is switched to solar power
source and the source selection is displayed on the LCD screen as shown in Fig. 9c.

During the off-peak time, from 20h00 to 07h00 the load is switched to the mains
power source and the source selection is displayed on the LCD screen as shown in
Fig. 9d.

When the mains is faulty at any time and it was selected, the load is switched to
solar power source and the source selection is displayed on the LCD screen as
shown in Fig. 10a.

When the solar power is faulty during the peak time, the load is switched to the
mains power and the selection is displayed on the LCD screen as shown in Fig. 10b

Fig. 9 a LCD screen showing solar power is selected during morning peak time. **b** LCD screen
showing mains power selected during off peak time. **c** LCD screen showing solar power selected
during evening peak time. **d** LCD screen showing mains power selected after peak time in the
evening

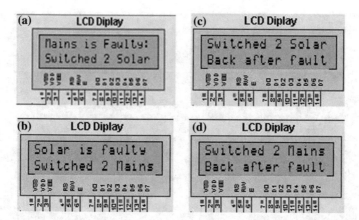

Fig. 10 **a** LCD screen selection of solar power when mains is faulty. **b** LCD screen showing selection of mains power when solar is faulty. **c** LCD screen showing selection of solar after fault condition is restored. **d** LCD scree showing returning to mains after fault condition is restored

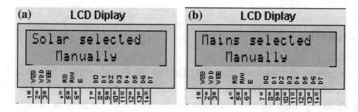

Fig. 11 **a** LCD screen showing manual selection of solar power. **b** LCD screen showing manual selection of mains power

When the solar power fault is restored during the peak time, the load is switched back to solar and the selection is displayed on the LCD screen as shown in Fig. 10c.

When the mains power fault is restored during the off-peak time, the load is switched back to mains and the selection is displayed on the LCD screen as shown in Fig. 10d.

When the user switches the load to the mains or solar power manually, the selection is displayed on the LCD screen as shown in Fig. 11a, b. The manual selection can override the peak time and availability of the sources in fuzzy programming.

The load analysis and energy consumed by different types of load is shown Fig. 12

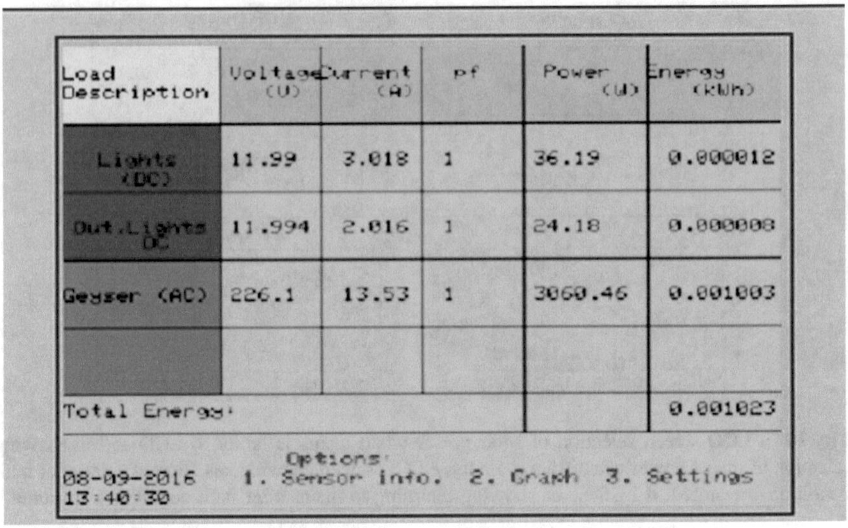

Load Description	Voltage (U)	Current (A)	pf	Power (U)	Energy (kWh)
Lights (DC)	11.99	3.018	1	36.19	0.000012
Out.Lights DC	11.994	2.016	1	24.18	0.000008
Geyser (AC)	226.1	13.53	1	3060.46	0.001003
Total Energy:					0.001023

Options:
08-09-2016   1. Sensor info.   2. Graph   3. Settings
13:40:30

Fig. 12 Load analysis for different residential loads

5 Conclusion

Peak time power demand causes fluctuations in the electrical network from the generation, transmission and distribution sections of the power grid. This instability globally causes problems of insufficiency in electricity access in remote areas. In this research, the hybrid system using solar and mains power as sources is designed. The fuzzy logic algorithm is used to select the power source based on the peak and off peak time, utilized power level and availability of power source to meet the power demand during the peak time and off peak time. The advantage of implementing the fuzzy logic system in selecting the power source based on the peak time, availability of the power source and manual input shows that the accuracy of selecting the source under different operating conditions. Also the energy consumed by different loads is shown for modification of the system for further analysis.

References

1. T. Gonen, *Electric Power Distribution Engineering* (2014)
2. C. Bayliss, C.R. Bayliss, B.J. Hardy, *Transmission and Distribution Electrical Engineering* (Elsevier, 2012)
3. S. Bhattacharya, *Basic Electrical and Electronics Engineering I (For Wbut)* (Pearson Education India, 2010)
4. C. Philibert, Case study 1: concentrating solar power technologies, in *International Energy Technology Collaboration and Climate Change Mitigation* (ed: OECD Environmental Directorate, IEA, Paris, 2004)

5. L. Schipper, S. Meyers, *Energy Efficiency and Human Activity: Past Trends, Future Prospects* (Cambridge University Press, USA, 1992)
6. T. Esram, P.L. Chapman, Comparison of photovoltaic array maximum power point tracking techniques. IEEE Trans. Energy Convers. EC **22**, 439 (2007)
7. P. Yeung, *Reducing Energy Costs with Peak Shaving in Industrial Environments* (2007). http://www.schneider-electric.com.hk/documents/energy-efficiency-forum/Reducing-Energy-Costs-with-Peak-Shaving.pdf
8. M.A. McNeil, V.E. Letschert, *Future Air Conditioning Energy Consumption in Developing Countries and What can be Done About it: The Potential of Efficiency in the Residential Sector* (Lawrence Berkeley National Laboratory, 2008)
9. D. Devos, Eskom's dim understanding of the market. Mail Guardian [Opinion] (2013), http://mg.co.za/article/2013-05-03-00-eskoms-dim-understanding-of-the-market
10. R.A. Schneider, G.L. Freed, R. Lipnick, *Power Monitoring and Load Shedding System* (ed: Google Patents, 1978)
11. P. Nema, R. Nema, S. Rangnekar, A current and future state of art development of hybrid energy system using wind and PV-solar: a review. Renew. Sustain. Energy Rev. **13**, 2096–2103 (2009)
12. J. Ding, J. Buckeridge, Design considerations for a sustainable hybrid energy system (2000)
13. S.J. Ehnberg, M.H. Bollen, Reliability of a small power system using solar power and hydro. Electr. Power Syst. Res. **74**, 119–127 (2005)
14. E. Koutroulis, K. Kalaitzakis, N.C. Voulgaris, Development of a microcontroller-based, photovoltaic maximum power point tracking control system. IEEE Trans. Power Electron. **16**, 46–54 (2001)
15. N. Ahmed, M. Miyatake, A stand-alone hybrid generation system combining solar photovoltaic and wind turbine with simple maximum power point tracking control, in *CES/IEEE 5th International Power Electronics and Motion Control Conference* (IPEMC 2006), pp. 1–7
16. G. Seeling-Hochmuth, A combined optimisation concept for the design and operation strategy of hybrid-PV energy systems. Sol. Energy **61**, 77–87 (1997)
17. T. Esram, P.L. Chapman, Comparison of photovoltaic array maximum power point tracking techniques. IEEE Trans. Energy Convers. EC **22**, 439 (2007)
18. R. Surtees, Electricity Demand Growth in South Africa and the Role of Demand Side Management, Eskom, PO Box, vol. 1091 (1998)
19. M. Habyarimana, V. Chitra, Automated hybrid solar and mains system for peak time power demand, in *23rd International Conference on Domestic Use of Energy*, vol. 1 (2015) pp. 169–175
20. T.J. Ross, *Fuzzy Logic with Engineering Applications* (2nd Edition, Wiley, 2004)

Energy Saving Using Memorization: A Novel Energy Efficient and Fault Tolerant Cluster Tree Algorithm for WSN

S.S. Jaspal, Umang and Brijesh Kumar

Abstract Sensor networks monitor and provide an insight into the events occurring in physical environment. Since nodes in a wireless sensor network (WSN) are energy constrained and replenishment of resources is generally not feasible, continuous research efforts are directed towards ensuring conservation of energy, so that network longevity may be maximized. This paper discusses the working philosophies of various clustering, tree and cluster-tree protocols and their energy conserving practices. A summarization of cluster-tree protocols with their salient features, problems addressed and limitations is also provided. Next, a mathematical analysis of energy consumption in ideal cases of cluster and cluster tree based protocols is presented. It is observed that cluster head election is the costliest process in a WSN and there is a scope to optimize this process in order to save energy. Hence, this paper proposes Energy Saving Using Memorization (ESUM); a novel energy-efficient and fault tolerant algorithm for cluster-tree based routing, that uses energy conservation to enhance network longevity, by using saved results and avoiding re-elections after each round. ESUM also handles CH failures efficiently as it uses memorized results to replace failed cluster heads thus saving both time and energy. The algorithm uses fuzzy logic for localization of motes to account for the irregularities like fading and multipath communication in wireless communication.

S.S. Jaspal (✉)
Amity University, Gurgaon, Haryana, India
e-mail: shalini.jaspal@gmail.com

Umang
Institute of Technology and Science, Mohan Nagar, Ghaziabad, UP, India
e-mail: singh.umang@rediffmail.com

B. Kumar
Nano Science and Technology, Amity School of Engineering and Technology,
Amity University, Gurgaon, India
e-mail: bkumar2@ggn.amity.edu

© Springer International Publishing AG 2017
A.K. Sangaiah et al. (eds.), *Intelligent Decision Support Systems*
for Sustainable Computing, Studies in Computational Intelligence 705,
DOI 10.1007/978-3-319-53153-3_10

Keywords WSN · Cluster based protocols · Tree based protocols · Cluster-tree based protocols · Energy conservation · Network longevity · Fault tolerance · Fuzzy logic · Localization

1   Introduction

Basic purpose behind setting up a WSN is to sense desired physical attributes of a region and communicate them back to the Base Station(s) [1]. This dynamic sensing and reporting has made WSNs penetrate every application that needs to respond to events in the real world. Use cases of WSNs range from smart homes, smart cities, smart agriculture, industrial monitoring and control, sensing military fields, detecting forest fires, tracking items in logistics to name a few. Some real life case studies where communities have benefited from the use of sensor networks include Smart Agriculture in Columbia that lead to improvement in production of banana crops, reduction of emissions in Nordic Smart Cities [2], the Zebranet project for wild-life tracking by Princeton University [3], use of wearable sensors for marketing, rehabilitation, sports coaching and many more purposes [4]. Considering such a wide range of application areas, researchers are continuously investing efforts in order to improve efficiency of sensor networks.

Since these networks are largely composed of energy constrained sensor nodes operating in an unattended mode, energy replenishment is generally not feasible. Communication in a WSN is the most energy demanding process thus minimization of communication overheads is an obvious means of conserving energy. Communication protocols for WSNs are largely dependent on the topology of the network which, in turn, decides the communication path [5] and hence the energy expended in the energy demanding task of communication. Over the years WSN protocols have evolved from simple and fault-tolerant direct communication protocols [6] to complex but energy conserving cluster-tree based communication protocols.

This section tracks the evolution of communication protocols in WSNs from the point of energy conservation and discusses the need and advantages of cluster-tree based routing.

1.1   *Energy Concerns of Older Communication Protocols in WSNs*

Direct Communication, the simplest communication protocol, required each node to sense the physical environment and communicate the generated data to the Base Station. The simplicity of direct communication made it fault tolerant since failure of a single node did not affect the rest of the network, but since every node in the network was involved in long distance communication, the total energy of the network depleted at a very fast rate, leading to an early death of the network [6].

This limitation paved the way for Cluster based networks, where the entire network was divided logically into clusters and the task of communication was limited to a small subset of nodes termed as the Cluster Heads (CHs). The member nodes (MNs) had the responsibility of sensing data and handing it over to a closely placed CH, responsible for communicating the data back to the base station (BS) [7]. Since the CHs were penalized nodes and would run out of energy very fast, variations in cluster based protocols allowed for rotating the role of CH amongst the member nodes. The decision of delegating the role was based on different parameters like placement of the node, its residual energy, number of times it had already played the role of CH etc. [8]. Still most protocols required the CH to aggregate and communicate the data directly to the BS in a single hop, thus draining out the energy of CHs and bringing down the total residual energy of the network.

Chain and tree based protocols aimed at saving energy by communicating the aggregated data to the base station in multiple hops [9], but as the number of hops increased so did the delay in communication [7, 8]. Also in case of tree based protocols, where every node of the network was a part of the tree, failure at any of the internal nodes resulted into loss of connectivity for all its descendants during the ongoing epoch i.e. till the tree was not reconstructed [8].

1.2 Advantage of Cluster-Tree Protocols

The cluster-tree topology, as per IEEE 802.15.4, is based on a tree structure formed by connecting cluster-heads. The cluster-tree scheme combines the advantages of both cluster and tree schemes, namely reduction in volume of transmitted data by means of aggregation as in clustered networks and multi-hop transmission as is done in case of tree based networks. The need for frequent restructuring, due to node failures as in case of tree based schemes, is also eliminated. Hence, cluster-tree based protocols are most advantageous from the perspective of energy conservation and enhancing network longevity.

Cluster-tree based protocols win by combining the merits of Cluster based and Tree based protocols, since they form a tree of CHs [10]. The cluster-tree network, like in case of cluster based protocols, is divided into sections termed as clusters and the CH plays the role of collecting and aggregating data from its cluster. After aggregation, this data is propagated to upper levels of the tree, till it finally reaches the BS. Cluster-tree networks offer a deterministic behavior thus making it easier to adhere to QoS standards [11]. The following section discusses two tree based protocols TREEPSI [12] TBDCS [13], which is followed by a similar discussion on cluster-tree based protocols and finally a summarized comparison of these protocols is presented in a tabulated form. This discussion is followed by presentation of the proposed protocol ESUM which helps to overcome the research gap in the other cluster-tree protocols. ESUM, like other protocols discussed in the text, at the moment assumes that all the nodes in the network are trustworthy. Later versions of ESUM though, will incorporate trust level calculation using direct and indirect trust functions as discussed in [14].

2 Related Work

2.1 TREEPSI [12]

TREEPSI works on a WSN constituted of randomly distributed nodes. It utilizes a tree like formation in order to gather and forward the data to the BS. The main objective behind the design of TREEPSI is to save energy and thus enhance network longevity. It records an improvement over LEACH [13] and PEGASIS [15] in terms of energy utilization, though the effect on delay involved in transfer of information is not documented by the researchers.

Working Philosophy: LEACH CHs consume a large amount of energy while transmitting aggregated data to the sink. This reduces the scalability of the protocol and makes it less suitable for a large network. PEGASIS on the other hand, either utilizes a NP-Complete or a greedy approach to form an optimal chain of all the nodes. TREEPSI improves on LEACH by using a tree to communicate with the sink thus obviating the need for one hop communication with the sink. Also the length of the path from the root to the sink is claimed to be the shortest, which makes it an improvement over PEGASIS.

Assumptions: The protocol works on basic assumptions that all the nodes in the network are homogenous, immobile, energy constrained and location aware. Also all sensor nodes are assumed to be capable of communicating with the sink and all other nodes in the network.

Proposed Concept: The tree is rooted at a random node and the construction of the tree may be done in two ways. The first way is to construct the tree at the base station and communicate the information of the path to all the nodes. The second way, on the other hand, runs a common algorithm on all the nodes and constructs the tree locally. After successful completion of the construction of tree structure, the information may either be pulled by the BS by using a control packet or may be pushed by leaf nodes towards the BS. The first approach may be utilized when the BS needs to query some specific information; the second approach on the other hand is suitable when an exceptional case is to be reported to the BS.

Advantage: The advantages of the protocol include a remarkable improvement (30% more) in energy efficiency as compared to PEGASIS and up to 300% better performance as compared to LEACH [12].

Issues/Limitations: Since a tree structure allows a single path between root and any other node of the tree, such protocols are sensitive to single node failures, as failure on any of the nodes makes the descendents of the node disconnected from the BS during the ongoing epoch [1]. Secondly, as listed in the algorithm, a node selected to be the root, continues to play the role, till it "has more than minimum necessary energy for transmitting to the base station". This fact implies that a node may play the role of root only once in its life time. An efficient choice of root would ensure that the node is close to the BS, since this would require less energy in communication. This implies nodes close to BS will soon run out of the energy, thus breaking the connectivity between the network and BS [16].

Future Scope: Enhance tolerance to single node failures and, as suggested by the authors, perform an analysis of TREEPSI with respect to delay in data collection and find suitable techniques for solving issues related to delay on transmission.

2.2   TBDCS [13]

Tree Based Data Collection Scheme for WSN uses tree topology, designed by minimizing intermediate nodes, to reduce network traffic and enhancing network longevity. Here, the intermediate nodes act as data aggregators which know about their next hop children (immediate descendants) and choose a proper time locally, to aggregate data.

The algorithm works by computing a Forwarding Node Set (FNS) using a greedy solution of the NP-hard vertex cover problem. A FNS tree is formed incrementally, starting from the sink, by computing the FNS for each node that receives the message sent by sink for forwarding user's request. The same tree is used for sending the data back to the BS. But, being a tree based protocol, it still has the limitation of sensitivity to single node failures.

2.3   GTC

Generic Top-down Cluster-tree algorithm utilizes the top down approach [17] for tree formation. This approach is chosen since it allows flexibility in cluster size, tree depth etc. by using parameters that control the characteristics of the clusters and the tree formation. Rather than depending on global location information, the algorithm uses the information of the topology, available locally with individual nodes and their neighbors, in order to construct the cluster-tree.

Working Philosophy: The process of cluster formation progresses in phases. It may either be initiated by one of the sensor nodes or by the Base Station. The initiating node broadcasts a message to begin formation of a cluster, the cluster in turn, is joined by all the nodes that get the message. The root node then identifies certain nodes close to the boundary and allots them the role of CHs for the next level. The process of formation of clusters is then delegated to the newly identified CHs and it continues till the entire sensed field is not covered. The CHs identified in the entire process follow a parent child relationship and form a tree of clusters thus serving as the backbone of the network. The size of the clusters is controlled by a parameter that decides the maximum number of hops allowed for a node to be chosen as a member of the cluster. The tree formation is followed by a self-optimization phase which further "reduces the depth of the cluster tree and improves routability".

Advantage: The properties of the network topology formed as a result of GTC, are comparable to the ideal Hexagonal Packing. The depth of the tree is further reduced after the self-optimization phase, which is run post cluster formation. Hence, the algorithm proves to be an efficient solution for organizing sensor nodes so that energy consumption is reduced and network longevity is enhanced [17, 18].

Problem Addressed: Uneven cluster size.

Issues/Limitations: Involvement of all nodes in re-election of CH, is costly in terms of time as well as energy [19].

2.4 OCTBR [20]

Optimized Clustering Tree Based Routing works on the assumption that nodes and the BS are static and CH communicates with MNs using TDMA. It also assumes that nodes belong to the same cluster throughout their lifetime. The protocol proceeds in two phases, cluster formation takes place in the first phase and the CHs form a cluster tree in conglomeration during the second phase.

Working Philosophy: During initialization, each node transmits its residual energy and its location to its neighbors, which in turn sends a feedback on receipt of the message. This helps the nodes to collect information related to their neighbors like their count, residual energy and distance. Nodes utilize this information to decide if they will play the role of CH or not. These CHs send messages to their neighbors and the non-CH nodes decide to join the nearest cluster. The formation of CH Tree is initiated by the BS, by sending a message to neighboring CHs, the CHs which receive this message join the Cluster Tree and the structure is further augmented by propagation if this messages to farther CHs.

Advantage: The protocol gives higher energy efficiency and hence enhanced network lifetime as compared to LEACH and PEGASIS.

Issues/Limitations: OCTBR involves all the nodes in re-election process, hence energy consumption is high. The algorithm also leads to uneven energy dissipation, since it doesn't aim to place the CH at center of the cluster.

2.5 CTDGA [10]

CTDGA is a Cluster Tree based algorithm that works by forming a tree of Cluster Head (CHs). This tree, rooted at the Base Station (BS), is used for the purpose of data dissemination.

Assumptions: The algorithm is based on the assumptions that the network is composed of homogenous, location unaware nodes that are distributed in a random fashion over the entire field.

Working Philosophy: The formation of the network structure, in CTDGA, is initiated by the Base Station by sending a broadcast packet, whose RSS is used by individual nodes to compute their distance from the BS. Every node of the network, then, broadcasts its residual energy, ID and its computed distance from BS. A table containing this broadcasted information, along with the state (CH/MN) is maintained by each node. Balanced Cluster formation is done using a function, where the number of member nodes in a cluster is proportional to the ratio of area of the cluster and the area of the whole network. Once the clusters are formed, all CHs are connected through a spanning tree and the data is transmitted towards the sink. Communication with the BS is done by a single CH termed as the Super CH. The role of Super CH is circulated between nodes using a function that is directly proportional to current energy of the node and inversely proportional to its distance from the sink.

Advantage: Simulation results show that network lifetime by using CTDGA is double of that obtained using LEACH.

Problems Addressed: CTDGA uses RSS for distance calculation, hence no extra hardware is required. Balanced Cluster formation is done using a function, where the number of member nodes in a cluster is proportional to the ratio of area of the cluster and the area of the whole network.

Issues/Limitations: Doesn't address node mobility. Since communication with sink is done through single Super CH, which may be located anywhere in the field, the choice would be energy demanding if the Super CH is far off from the sink.

2.6 CIDT [21]

Cluster Independent Data collection Tree, proposes usage of a Data Collection Tree (DCT) which is independent of the node clusters contained in the network. The algorithm aims at addressing the problem of node mobility which leads to frequent changes in the topology of the network.

Working Philosophy: The Cluster Head, chosen by the Member Nodes on basis of the RSS and connection time, is responsible for data collection (using TDMA), data aggregation and cluster maintenance. Communication of collected and aggregated data is done on a separate tree, termed as Data Collection Tree (DCT), which is composed of nodes other than Cluster Heads. Hence, unlike other Cluster-Tree based algorithms, in CIDT, CHs don't have the responsibility of communication which is a high energy consuming task. The parameters used for selecting Data Collection Nodes (DCN) include a threshold energy value, the received signal strength and the time for data transmission. The DCNs are elected by the Base Station and don't participate in sensing and also, don't belong to any specific cluster. DCNs are ensured to have a good connectivity amongst themselves and the CHs. The roles of DCNs include collecting data from CHs, aggregating them and forwarding the data (using DSSS) to the BS along the DCT.

The movement of sensor nodes doesn't impact the network structure as long as the CH or the DCNs don't move and the sensor nodes remain within the same cluster. High mobility of nodes, CH or DCN leads to restructuring of the network. Also, the DCT needs to be rebuilt in case the CHs change or get repositioned.

Advantage: CIDT leads to a much higher PDR and Throughput as compared to LEACH and HEED and also addresses issues of node mobility.

Problems Addressed: Node Mobility

Issues/Limitations: Requires re-election of DCN every time CHs are changed due to failure or mobility. Two elected entities involved.

2.7 VELCT [22]

"Velocity Energy-efficient and Link-aware Cluster Tree" is an enhancement over CIDT and alleviates various issues of CIDT like node mobility, coverage and reduces delay. VELCT provides stable links by reducing the communication overhead from the CHs and by selecting links with maximum connection time and received signal strength.

Advantage: Higher PDR, lesser delay and lesser energy consumed as compared to LEACH, HEED and CIDT.

Problems Addressed: Increased cluster longevity by choosing links with maximum connection time and RSS.

Issues/Limitations: Requires re-election of DCN every time CHs are changed due to failure or mobility.

Simulations on NS [21, 22] with 500 nodes in a region of $1000 \times 1000 \text{ m}^2$ having data packets sized 512 bytes, cluster size ~40 m, sensing range ~20 m, yielded the results tabulated (Table 1).

2.8 CTDD [23]

Cluster-tree based Data Dissemination, assumes static nodes and a mobile sink. CIDT proceeds by dividing the region to be sensed into a virtual grid having equal

Table 1 Simulation result based comparison of cluster and cluster-tree based protocols [21, 22]

	PDR %	Delay (ms)	Total energy consumed (mJ)
LEACH	79	10	230
HEED	84	9.5	220
CIDT	93	6.5	200
VELCT	98	3	125

sized cells. After the formation of grids, CHs are selected from each grid. The nodes which are located at the center of the grid and have high residual energy are chosen to be the CHs. Once the selection of CH is over, a tree of all the CHs is constructed. When the energy of CH falls below a threshold, node with maximal energy is chosen to be the new CH. The nodes at lower level accept this node as the new parent, when they receive a message from it, thus avoiding immediate reconstruction of the tree.

Advantage: Energy saving by conducting re-election only for low-energy CHs.

Problems Addressed: Need for energy conservation.

Concerns: Chooses CH from nodes located around the center of fixed set of grids, hence the role of CH is limited to a small set of nodes. Two elected entities involved.

2.9 CTRP [24]

CTRP is again a cluster tree protocol, with clusters based on a grid. The protocol works towards a mobile sink but still has the same limitation as CTDD, which tries to place CHs at the center of fixed grid formed during cluster field initialization.

3 Comparative Chart of Topology Building Approaches in WSN

Based on the preceding discussion, a summarized comparison of topology building protocols in WSN, is presented in Table 2.

4 Research Gap

Constrained energy and bandwidth, requirement to work with limited hardware in unattended mode, need to know node location and limited scalability of networks [24] are some of the challenges faced by WSNs. Communication being the most energy demanding task, algorithms that can help in reducing the bulk of messages exchanged are the need of the hour.

In has been observed from the discussion in previous section that cluster-tree protocols outperform their counterparts by conserving energy and enhancing network longevity. But, all the discussed approaches of cluster-tree formation resort to immediate re-election of CH, after each epoch. Similar action is taken in case of failure/movement of the cluster head. Custer Head election involves a huge communication overhead. Algorithms that build cluster-trees, involve multiple message

Table 2 Comparison of leading cluster, chain, tree and cluster-tree based protocols

Protocol	N/W architecture	Nature of nodes	Salient features	Problem(s) addressed	Issues
LEACH (2000)	Cluster based	Homogeneous, energy constrained, immobile nodes	Minimizes total energy used by rotating the role of CH	Increase in network lifetime by 8 × as compared to direct-transmission	One hop connectivity between CH and BS
PEGASIS (2002)	Chain based	Homogeneous, energy constrained, immobile nodes	Uses chain based forwarding to BS	Saves up to 300% more energy as compared to LEACH by eliminating the need for re-election of CHs	Delay in transmission
TREEPSI (2006)	Tree based	Homogeneous, energy constrained immobile and location aware nodes	Tree based connectivity decreases the distance over which data is to be communicated by individual nodes	Increased life time as compared to cluster based protocols which depend on single hop communication with the sink	Prone to faults in case of single node failures Increases delay
TBDCS (2006)	Tree based	Static nodes that know their hop distance to sink, their one-hop neighbors and their remaining energy	Builds FNS tree, by using the vertex-cover problem	Increased lifetime as compared to LEACH without need for global information	Prone to faults in case of single node failures Increases delay
GTC (2011)	Cluster tree	Large scale WSN. No requirement of neighborhood information or location awareness	Parameter based control on network characteristics Optimization phase that gives near to hexagonally packed clusters	Uneven cluster size and holes in sensed region	Re-election costly in terms of time and energy
OCTBR (2012)	Cluster tree	Homogeneous, immobile nodes that belong to same cluster always	CH decided on basis of neighbor count, residual energy and distance	More energy efficiency and network lifetime as compared to LEACH	Doesn't aim to place CH at center, hence uneven energy dissipation

(continued)

Table 2 (continued)

Protocol	N/W architecture	Nature of nodes	Salient features	Problem(s) addressed	Issues
CTDGA (2012)	Cluster tree	Homogeneous, immobile nodes that find their location using RSSI of signal broadcast from BS	Number of member nodes in a cluster is proportional to the ratio of area of the cluster and the area of the whole network	Enhancement of network lifetime as compared to LEACH	Nodes broadcast energy and location information Communication with sink done by a Super CH in a single hop
CIDT (2014)	Cluster tree	Densely populated WSN with homogeneous nodes	A separate DCT whose nodes are not a part of any specific cluster	Node mobility	Failure and mobility of cluster head Requires DCN re-election with change in CHs
VELCT (2015)	Cluster tree	Densely populated WSN with homogeneous nodes	A separate DCT whose nodes are not a part of any specific cluster	Node mobility is addressed better by selecting stable links	Requires DCN re-election with change in CHs
CTDD (2015)	Cluster tree	Densely populated WSN with homogeneous nodes	Tries to avoid immediate reconstruction of tree by replacing only those cluster head which are low on energy	Energy saving Increased reliability as compared to tree based communication	The role of CH is limited to a small set of nodes

exchanges between all the potential cluster heads and the rest of the network, after each round. This process may be optimized in order to save energy.

CH re-election is a major energy consuming process in any clustering protocol, and CTDD [23] attempts to avoid it by replacing only those CHs which are low on energy, but in CTDD CHs are always placed near the center of pre-decided fixed cluster. This implies that only a limited set of nodes get to become the cluster heads, thus leading to uneven energy dissipation.

The next section of the chapter presents an alternate approach, ESUM, which addresses the issue by using memorization to avoid immediate re-election and thus leads to energy conservation and enhanced network lifetime. The algorithm also attempts that the CH role is assigned to all the nodes in a systematic fashion so that the residual energy is evenly distributed in the entire network, thus reducing the chances of network partitioning.

5   Proposed Solution ESUM

The discussion presents Energy Saving Using Memorization (ESUM), a novel energy-efficient and fault tolerant algorithm for cluster-tree based routing, that works towards increasing the network lifetime. ESUM also increases fault tolerance as it uses memorized results to replace failed cluster heads thus saving both time and energy. ESUM relies on two facts; firstly sensor nodes in terrestrial WSNs are immobile in nature and secondly large distance communication demands much more energy as compared to the combined energy requirements of sensing, local data processing and short distance communication.

The protocol proceeds in phases. The first phase starts from the BS, which along with the CHs chosen at each subsequent phase, plays the role of Topology Augmenter. To begin with, the BS builds a cluster around itself, and polls the nodes at a distance equal to desired cluster size (CS), for readiness to play the role of CH at the next level. The nodes that receive the message decide to respond or not, on basis of their residual energies and a flag pchState. This set of responding nodes is maintained as Potential Child Cluster Heads (PCCH) by the Topology Augmenter for the current cycle. The PCCH set is used for deciding the Active Cluster Head of the child clusters after fixed time periods, removing the node that has played the role of CH from the PCCH, with each selection. Since, at any instance of time, the members of PCCH have not acted as the CH in the current cycle, these are expected to be rich in energy, though an additional check of current energy level may be made before delegation of responsibilities. The number of re-elections avoided is, in the best case, equal to the count of members of the PCCH. The process of rebuilding the PCCH begins after the current set has been exhausted, and this marks the beginning of a new cycle. The position of first level PCCH is changed after each cycle, in order to give all nodes an equal opportunity to bear the responsibility of heading a cluster, thus leading to even consumption of energy in the network. ESUM uses weighted centroid scheme for localization, where weights are calculated using Fuzzy Inference System in order to incorporate irregularities in real life environments.

Data maintained by each Topology Augmenter for building and maintaining child clusters, includes:

1. Potential Child Cluster Heads (PCCHs) which are set of high energy nodes at a given instance of time, and may serve as CH for child clusters in the forthcoming rounds.
2. Active Child Cluster Heads (ACCH) which is the set of Cluster Heads of child clusters, active at a given instance of time.
3. pchState which is maintained by each node to indicate if it is a potential cluster head. The state is initially set, and is unset for each node once it has played the CH role. A node whose pchState is unset is not chosen to head a cluster, since it has already utilized a large amount of energy.

5.1 ESUM Algorithm

Case I: When PCCH is empty:
Activity: Discover PCCH (Refer Annexure 2, Figs. 8 and 9)

1. The cluster formation process is initiated by the BS by sending CH_REQ signals. The BS also initiates a timer, whose timeout event triggers the rotation of CH role.
2. The nodes receiving these signals respond by CH_RES, provided their pchState is set and the residual energy is above threshold. The responding nodes compute their location information using weighted centroid method, where weights are computed using Fuzzy Inference Scheme.
3. All such CH_RES received by the BS, are remembered at the BS. The information regarding responding nodes is used to form the set termed as Potential Child Cluster Heads (PCCH).

Activity: Build ACCH (Refer Annexure 2, Fig. 10)

4. The CHs for the current round are selected out of the PCCH, thus forming the ACCH i.e. CHs at next level of the tree. The current energy levels and location information of members of PCCH is used to ensure healthy cluster heads and evenly sized child clusters. At this stage the pchState of the new ACCH members is unset and these nodes are removed from the PCCH.

Activity: Build Cluster

5. The BS also sends a M_REQ. The M_REQ is meant for member nodes in vicinity of BS, which in turn respond by M_RES to join the cluster.

The process of sending CH_REQ and M_REQ for building the cluster-tree is continued by all members of ACCHs at each level, thus augmenting the network topology. Topology augmentation continues till the region boundary is not reached, thus leading to a cluster tree that spans the entire region.

Case II: When timer has expired and PCCH is not empty: -
Activity: Build ACCH (Refer Annexure 2, Fig. 10)

When Timer initiated by the BS expires, the BS starts the process of rediscovery of ACCH for the next round. The new ACCH is formed in phases, level by level, as was done in the first round. But here, instead of exchange of CH_REQ & CH_RES, the existing PCCH is used to compute the new ACCH. Each Topology Augmenter at Level X (starting from the Base Station) performs the following steps for each ACCH at Level X + 1

6. Poll next PCCH member at level X + 1 for availability (which will succeed in most cases as the PCCHs have not yet played the role of CH and hence have sufficient energy)
7. If the polled PCCH member responds positively, give the role of ACCH to the responding member of PCCH and remove it from the set PCCH.
8. Else, unset the pchState flag, remove the polled node from PCCH and poll the next PCCH member. If at any stage PCCH is empty, go to Discover PCCH.

Case III: When timer has expired and PCCH is empty: -
Activity: Discover PCCH + Build ACCH

In case Topology Augmenter at Level X senses that PCCH at Level X + 1 is empty, it sends CH_REQ signals in order to rebuild the PCCH i.e. Case I.

5.2 Energy Efficiency

The stated approach maintains a list of potential cluster heads to ensure that the high energy consuming process of electing CHs is postponed for several rounds till the PCCH is not empty. Hence, the overhead of energy expenditure is reduced, without the need for immediate re-elections thus leading to energy conservation and enhancement in network longevity. Also, in case of CH failure, instead of resorting to immediate re-election, an appropriate member of PCCH is chosen by the parent CH, to replace the child CH that failed.

5.3 Fault Tolerance

Once the cluster-tree is completely formed, each CH knows about the PCCHs for all of its child clusters. In case an ACCH member fails, the parent CH shifts the responsibility of heading the corresponding child cluster to a PCCH member that belongs to the corresponding child cluster and is closest to the failed ACCH member.

5.4   Comparison of Energy Requirements After Initial Topology Setup

ESUM versus Cluster Based Protocols: Considering the results depicted in [25], it is known that LEACH exhibits best performance if number of Cluster Heads is between 4–5% of the deployed nodes.

Figures 1 and 2 depict the ideal scenarios on a 200×200 m^2 region, with 400 deployed nodes, for distribution of cluster heads using any clustering algorithm and ESUM respectively. The scenario assumes evenly formed clusters with cluster heads placed at the center of the cluster. In Fig. 2 the distance between two CHs within the field is CS, whereas the distance between BS and next level CHs is CS/2 in order to reduce the energy consumption in level I CHs.

As per the first order radio model [26] the energy consumed in transmitting one bit over a distance d is given by the formula:

$$E_{Tx} = E_{elec} + \varepsilon_{amp} * d^2 \tag{1}$$

Generalizing the case of ideal cluster based scenario (approaching $N \times N$ grid like structure), the total energy consumed in energy dissemination is given by the relation:

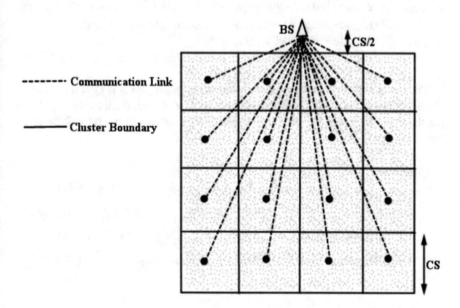

Fig. 1 Ideal scenario for evenly built clusters (close to a grid, single hop communication)

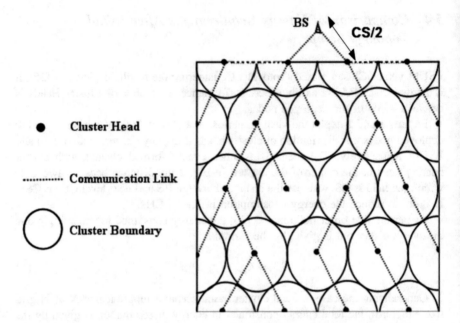

Fig. 2 Ideal scenario for ESUM (Multi-hop communication), member nodes are not depicted

$$E_{TotTx\_Clust} = (N^2 * E_{elec}) + 2 * \left[N/2 * (1^2 + 2^2 + 3^2 + \ldots + N^2) + N/4 * (1^2 + 3^2 + 5^2 + \ldots + (N-1)^2 \right] * \varepsilon_{amp} * CS^2$$

$$E_{TotTx\_Clust} = (N^2 * E_{elec}) + 2 * [N/2(N)(N+1)(2N+1)/6 + N/4(N/2)(N-1)(N+1)/3] * \varepsilon_{amp} * CS^2$$

$$E_{TotTx\_Clust} = (N^2 * E_{elec}) + ((5N^4 + 6N^3 + N^2)/12) * \varepsilon_{amp} * CS^2$$

$$(2)$$

Thus, it is evident that the energy consumed in transmission is $O(N^4)$ for a N × N grid. Solving the same for a specific case of 4 × 4 grid, depicted in Fig. 1. As per the depicted topology, the transmission energy consumed per bit will be given by:

$$
\begin{aligned}
E_{TotTx\_Clust} \\
&= 2 * (E_{elec} + \varepsilon_{amp} * CS^2 + (CS/2)^2) + 2 * (E_{elec} + \varepsilon_{amp} * CS^2 + (3CS/2)^2) \\
&+ 2 * (E_{elec} + \varepsilon_{amp} * (2CS)^2 + (CS/2)^2) + 2 * (E_{elec} + \varepsilon_{amp} * (2CS)^2 + (3CS/2)^2) \\
&+ 2 * (E_{elec} + \varepsilon_{amp} * (3CS)^2 + (CS/2)^2) + 2 * (E_{elec} + \varepsilon_{amp} * (3CS)^2 + (3CS/2)^2) \\
&+ 2 * (E_{elec} + \varepsilon_{amp} * (4CS)^2 + (CS/2)^2) + 2 * (E_{elec} + \varepsilon_{amp} * (4CS)^2 + (3CS/2)^2)
\end{aligned}
$$

Table 3 Energy utilization for data dissemination, in ESUM example scenario

S. no.	Node count	Packets trans.	Trans. dist.	E_{Tx}	Packets recd	E_{Rx}
1	10	1	CS	$10 * E_{elec} + \varepsilon_{amp} * 10 * P * CS^2$	0	0
2	5	2	CS	$5 * E_{elec} + \varepsilon_{amp} * 10 * P * CS^2$	1	$5 * 1 * E_{elec}$
3	1	3	CS	$E_{elec} + \varepsilon_{amp} * 3 * P * CS^2$	2	$1 * 2 * E_{elec}$
4	2	5	CS	$2 * E_{elec} + \varepsilon_{amp} * 10 * P * CS^2$	4	$2 * 4 * E_{elec}$
5	1	7	CS	$E_{elec} + \varepsilon_{amp} * 7 * P * CS^2$	6	$1 * 6 * E_{elec}$
6	1	8	CS	$E_{elec} + \varepsilon_{amp} * 8 * P * CS^2$	7	$1 * 7 * E_{elec}$
7	2	11	CS/2	$2 * E_{elec} + \varepsilon_{amp} * 11/2 * P * CS^2$	10	$2 * 10 * E_{elec}$

On simplifying, we get results in accordance with Eq. (2), given by:

$$E_{TotTx\_Clust} = 16E_{elec} + 140\varepsilon_{amp} * CS^2 \tag{3}$$

ESUM, on the other hand uses multi-level hierarchy to deliver sensed data to the Base Station, where each CH transmits gathered data to its parent CH. Hence, in each case the distance over which data is transmitted remains equal to the cluster size (CS). Although in case of cluster-tree topology, there is an additional overhead of receiving aggregated information from lower level CHs, which is given by E_{elec} per received bit; this component is much lesser in comparison to the energy required for transmission [26]. The total energy for dissemination in ESUM, considering the fact that that information sensed at level L travels L − 1 links to reach the BS, is given by:

$$E_{TotESUM} = E_{Tx\_ESUM} + E_{Rx\_ESUM} \tag{4}$$

Using Table 3, total energy required for transmitting a single bit over distance d, on the ideal topology of ESUM, the total energy consumed per bit by ESUM is given by:

$$E_{TotTx\_ESUM} \approx 70 * E_{elec} + 54\varepsilon_{amp} * CS^2 \tag{5}$$

Since $E_{elec} \ll \varepsilon_{amp}$ [26], this is a major improvement over the ideal clustering scenario. It is important to note that even though the number of clusters formed in ESUM is greater than the number of clusters in simple clustering, still the energy expended in transmitting sensed data to the CH would be lesser, because the total number of nodes involved in sensing the region is the same in both cases; but in ESUM, the smaller cluster size reduces the average distance between the MNs and the CHs (Table 3).

5.5   Comparison of Energy Requirements for Re-election

Various clustering and cluster-tree based protocols discussed in the foregoing text, involved all the nodes in the re-election process which was carried out after each epoch. This universal involvement implies interchange of messages on a very large scale and thus leads to heavy energy drainage [23]. Though CTDD tries to limit re-election to energy constrained CHs, still it involves all the member nodes in re-election process. ESUM on the other hand skips few election rounds by utilizing the saved results of previous elections.

In ESUM, once a member of PCCH has played the CH role, its pchState is unset and the node is removed from the PCCH set. After exhaustion of the PCCH sets, the base station re-initiates the process of building the new PCCH sets, starting from a distance CS/2+ cycle Number, thus building a new PCCH after every cycle; and therefore each node of the network gets an opportunity to head clusters. This approach ensures uniform load balancing, hence increasing network longevity Figs. 3 and 4 depict the selection of Child Cluster Head (CCH) from the PCCH set and the formation of clusters around each member of ACCH respectively.

5.6   Node Localization in WSN

Localization is important for sensor networks, since knowledge of position of motes makes it possible to find out the location of occurrence of an exceptional event, which is one of important use cases of a sensor field. In case of cluster based information gathering, location information of motes may be used to build evenly spread clusters, thus leading to efficient dissemination of data. Working of ESUM has a high level of dependency on computation of position of motes in order to build an effective PCCH. Hence, a detailed discussion on choice of localization approach is presented.

Different approaches to localization include utilization of a GPS module; and calculation of location using range based or range free methods. All the mentioned approaches have their advantages and disadvantages and these need to be kept in mind while choosing the localization approach to be followed in a particular implementation.

Inclusion of GPS module provides precise global positioning information, but the GPS system is power hungry, has problems in miniaturization and is generally inefficient in areas which are impenetrable by satellite signals [27].

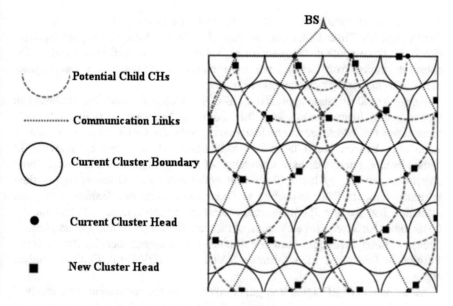

Fig. 3 Depicting selection of CCH from the PCCH set

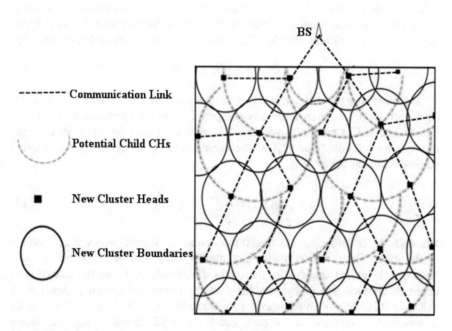

Fig. 4 Depicting formation of clusters around new CCH

Range based localization schemes depend strictly on metrics like Angle of Arrival (AoA), Time Difference of Arrival (TDoA), Received Signal Strength (RSS) etc. to compute distance and/or angle between individual sensor nodes. The approach, though very precise, poses a problem of additional hardware requirements and high energy consumption [28, 29].

Range free localization approaches [28–30] rely on radio modules, present in all sensor nodes, hence do not levy the additional hardware cost, like ultrasonic transceivers etc. which need to be integrated specifically for the purpose of location estimation. Beacon based range-free localization schemes rely on presence of certain beacon nodes that have knowledge of their position in the network. These algorithms proceed by letting the sensor nodes that have no knowledge of their position, compute their respective positions in a distributed fashion, using the estimated distance in a 2D plane from at least three beacon nodes. 3D positioning, on the other hand, requires estimation of distance from four, position aware, beacon nodes. Various techniques used for distance measurement include, Approximate Position in Triangle (APIT), Distance Vector—Hop (DV-Hop), Centroid method etc.

Measurement of RF signal strength in mW is often not convenient since received power P_r is inversely proportional to the d^n, the exponent n being dependent on the propagation environment. Because of the exponential relation, small changes in distance lead to significant changes in received power levels. The dBm scale, which is logarithmic in nature, is used to represent RSSI in order to overcome this issue. The received power, when measured in dBm, varies linearly with distance, hence making the computations simpler. The formula for conversion is defined as [31]:

$$dBm = \log(mW) * 10 \tag{6}$$

Assuming a sensor field, where sensor nodes and location aware beacon nodes have been set up; a sensor node may compute its position by communicating with beacon nodes. If a mote receives signal from at least three beacon nodes, it may estimate its position (x_{est}, y_{est}) in the sensor field by using the basic Centriod Method as given below [32]:

$$(x_{est}, y_{est}) = \left(\sum_{i=1}^{n} x_i, \sum_{i=1}^{n} y_i \right) \tag{7}$$

where (x_i, y_i) is the location of the ith location aware anchor node which is visible to the mote. Here, the term visibility implies capability of reception of radio messages. Now, the strengths of received radio signals may vary because of variations in distance and topography between the motes and anchor nodes. Lower values of RSSI from a given beacon, may either be caused by larger distance or by presence of an obstacle in the path, but higher RSSI definitely signifies closer placement and clear visibility. Hence, the basic centroid approach gives better results if weights are assigned to each anchor node on basis of RSSI value, giving higher weight to the anchor node from which stronger signal is received. Thus, the position of a mote may now be computed using the modified formula that accommodates these weights [33]:

$$(x_{est}, y_{est}) = \left(\sum_{i=1}^{n} w_i x_i / \sum_{i=1}^{n} w_i, \ \sum_{i=1}^{n} w_i y_i / \sum_{i=1}^{n} w_i \right) \tag{8}$$

5.7 Using Fuzzy Inference System to Overcome Problems of Crisp Programming

Assignment of weights on basis of RSSI is a complicated process since it is not possible to associate a crisp logic to compute RSSI on basis of distance. The mathematical formulae that relate RSSI to distance assume free space LOS communication. This approach faces inherent problems due to the fact that ideal line of sight, free space conditions are very unlikely to exist. Thus, issues in wireless communication like fading, multipath propagation etc. [33], make sole dependency on crisp logic inadequate.

In order to account for uncertainties in real world situations, Fuzzy Inference System (FIS) based on rules presented in Table 4, may be utilized to map a variant range of input values of RSSI, to a definite value of weight. As depicted in the simulation and experimental results presented by researchers [33–35], utilization of fuzzy logic in solving the localization problem, leads to a reduced error percentage as compared to the results obtained by pure mathematical model.

Fuzzy Logic helps in inferring crisp results utilizing overlapping and noisy inputs. Considering the example specification given in [36], the nodes don't respond to RSSI falling below −80 dB whereas a signal above −40 dB is considered to be equivalent to full visibility. Hence, the Input Membership Function for RSSI may be plotted as trapezoidal functions associated with five input ranges (Exceptional, Very Good, Good, Marginal and Intermittent) respectively and the Output Membership Function for weight in the range [0, 1] may be plotted as given (Figs. 5, 6 and 7).

The results of Mamdani FIS may be used to obtain value of weight for a particular value of RSSI. Since, the usage of weighted centroid method that uses

Table 4 Fuzzy logic rules to associate input RSSI to output weight

S. no.	Input (RSSI)	Output (weight)
1.	Intermittent	Very less
2.	Marginal	Less
3.	Good	Medium
4.	Very good	Large
5.	Exceptional	Very large

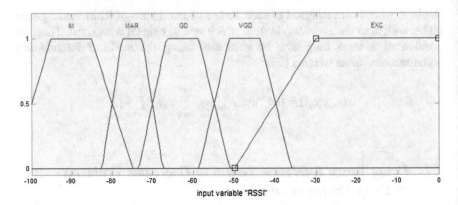

Fig. 5 FIS input variable RSSI

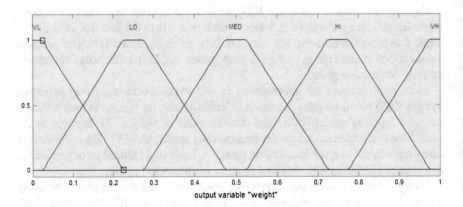

Fig. 6 FIS output variable weight

FIS based weight computation, has been proven to give better localization results as compared to the basic centroid method, implementation of ESUM would benefit from the approach as it is dependent on the position of nodes while forming the PCCH set.

6 Advantages of ESUM

Table 5 compares the bulk of messages exchanged, while choosing a new CH and replacing a failed CH, for ESUM and re-election based Cluster Tree protocols. It is clearly visible that ESUM has lower communication overhead than its counterparts

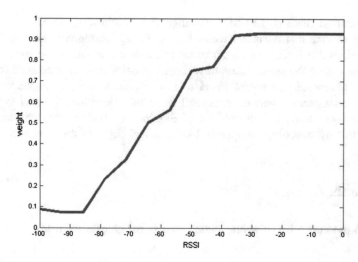

Fig. 7 Results of weight versus RSSI as generated by MATLAB

Table 5 ESUM versus re-election based cluster tree protocols

Activity	Message exchanges (ESUM)	Message exchanges (Re-election)	Advantage
Choosing new CH	Query and response messages between parent CH and a single member of PCCH	Message exchange between all nodes that are potential CHs	Reduction in communication overhead
Replacing failed CH	Query and response messages between parent CH and a single member of PCCH	Message exchange between all nodes that are potential CHs	Reduction in communication overhead

for both the listed activities. Further, since ESUM uniquely identifies the nodes in the governed network, it is suitable for applications that use node identifiers for purposes like dynamic configuration of nodes and spotting malicious behavior, intrusion detection etc.

7 Conclusion

The discussion presents a comparison of various cluster-tree protocols and proposes a novel algorithm ESUM that leads to increase in network lifetime by reducing the communication overhead in CH reelection. ESUM is dependent on location information, hence in order to give better localization results as compared to the basic centroid method, ESUM uses range free localization based on weighted centriod scheme that computes weights using Fuzzy Inference System so that the irregularities like fading, multipath communication etc. in wireless communication that arise due to the shear nature of real life scenarios may be accounted for.

The concept of cluster-tree based routing applies to densely populated, large scale, wireless sensor networks; hence ESUM may be efficiently used to govern such networks only. Future versions of ESUM may explore optimization of the cluster tree by controlling properties like span and depth of the sub-trees.

Appendix A

Table 6 presents a list of abbreviations used in the chapter for ready reference.

Table 6 Table of abbreviations

Abbreviation	Meaning
WSN	Wireless sensor network
BS	Base station
CH	Cluster head
MN	Member node
RSS	Received signal strength
AoA	Angle of arrival
TDoA	Time difference of arrival
APIT	Approximate position in triangle
RF	Radio frequency
RSSI	Received signal strength indicator
FIS	Fuzzy inference system
LOS	Line of sight
DSSS	Direct-sequence spread spectrum
PDR	Packet delivery ratio
CS	Cluster size
ESUM	Energy saving through memorization
PCCH	Potential child cluster head
ACCH	Active child cluster head
CH_REQ	CH request
CH_RES	CH response
M_REQ	Member request
M_RES	Member response

Appendix B: Activity Diagrams

Figures 8, 9 and 10 present the activity diagrams for important modules of ESUM algorithm.

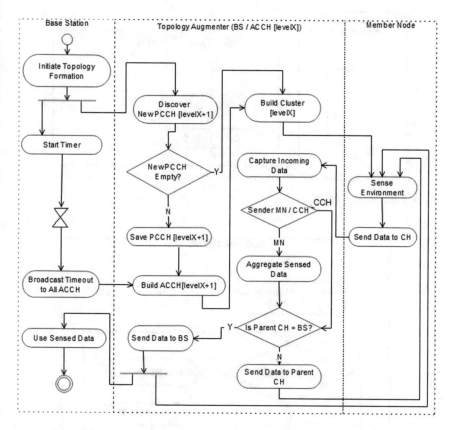

Fig. 8 Activity diagram representing Topology Augmentation and one cycle of retrieving sensed data

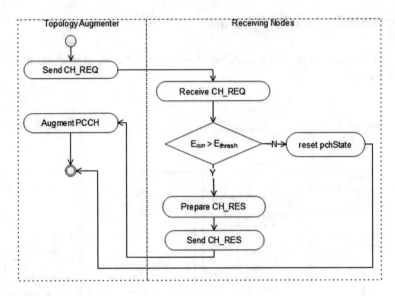

Fig. 9 Discovering a PCCH member

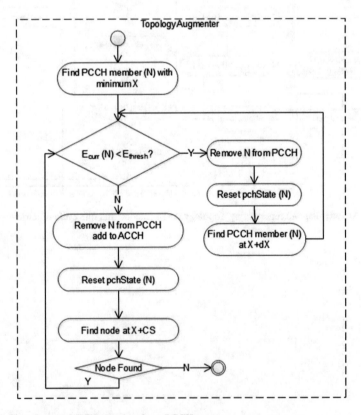

Fig. 10 Choosing an ACCH member from PCCH set

References

1. J. Elson, D. Estrin, Wireless Sensor Networks: A Bridge to the Physical World, in *Wireless Sensor Networks* (Springer, New York, 2004)
2. http://www.libelium.com/libeliumworld/case-studies/. Accessed 10 Nov 2016
3. M. Martonosi, *The Princeton ZebraNet Project: Sensor Networks for Wildlife Tracking* (Department of Electrical Engineering, Princeton University)
4. http://www.shimmersensing.com/research-and-education/current-use/wsn-case-study/. Accessed 10 Nov 2016
5. H. Karl, A. Willig, Protocols and Architectures for Wireless Sensor Networks, in *Routing Protocols* (Wiley Publications, England, 2005)
6. Kemal Akkaya, Mohamed Younis, A survey on routing protocols for wireless sensor networks. Ad Hoc Netw. **3**(3), 325–349 (2005)
7. Xuxun Liu, A survey on clustering routing protocols in wireless sensor networks. Sensors **2012**(12), 11113–11153 (2012). doi:10.3390/s120811113
8. Quazi Mamun, A qualitative comparison of different logical topologies for wireless sensor networks. Sensors **2012**(12), 14887–14913 (2012). doi:10.3390/s121114887
9. Z. Zhang, F. Yuv, Performance analysis of cluster-based and tree based routing protocols for wireless sensor networks, in *Proceedings of International Conference on Commun. Mobile Comput. (CMC)*, Shenzhen, China, April 2010, vol. 1 (2010), p. 418
10. J. Yang, B. Bai, H. Li, A cluster-tree based data gathering algorithm for wireless sensor networks, in *IEEE International Conference Automatic Control and Artificial Intelligence (ACAI 2012)* (2012). doi:10.1049/cp.2012.0910
11. Ricardo Severino, Nuno Pereira, Eduardo Tovar, Dynamic cluster scheduling for cluster-tree WSNs. SpringerPlus **3**, 493 (2014). doi:10.1186/2193-1801-3-493
12. S.S. Satapathy, N. Sarma, TREEPSI: tree based energy efficient protocol for sensor information, in *Proceedings of International Conference on Wireless and Optical Communications Networks (IFIP)*, Bangalore, India (2006), p. 4
13. H. Li, H. Yu, A. Liu, A tree based data collection scheme for wireless sensor network, in *Proceedings of International Conference on Networking, International Conference on Systems, International Conference on Mobile Communications and Learning Technologies (ICNICONSMCL)*, Morne, Mauritius, April 2006 (2006), p 119
14. S. Umang, B.V.R. Reddy, M.N. Hoda, Enhanced intrusion detection system for malicious node detection in ad hoc routing protocols using minimal energy consumption. IET Commun. **4**(17) (2010). doi:10.1049/iet-com.2009.0616
15. S. Lindsey, C.S. Raghavendra, PEGASIS: power efficient gathering in sensor information systems, in *Proceedings of the IEEE Aerospace Conference*, Big Sky, Montana, March 2002 (2002)
16. J. Li, P. Mohapatra (2007) Analytical modeling and mitigation techniques for the energy-hole problem in sensor networks. Pervasive Mob. Comput. 233–254 (2007)
17. H.M.N.D. Bandara, A.P. Jayasumana, T.H. Illangasekare, A top-down clustering and cluster-tree-based routing scheme for wireless sensor networks. Int. J. Distrib. Sens. Netw. (2011). doi:10.1155/2011/940751 (Hindawi Publishing Corporation)
18. H.M.N.D. Bandara, Top-Down Clustering Based Self-organization of Collaborative Wireless Sensor Networks. MS Thesis 2008, Department of Electrical and Computer Engineering, Colorado State University, Colorado (2008)
19. A.E.A.A. Abdulla, H. Nishiyama, N. Ansari, N. Kato, Energy-aware routing for wireless sensor networks, in *The Art of Wireless Sensor Networks: Volume 1-Fundamentals* (Springer Science & Business Media, 2014)
20. J. Zhang, Y. Xie, D. Liu, Z. Zhang, OCTBR: optimized clustering tree based routing protocol for wireless sensor networks, in *Internet of Things Volume 312 of the series Communications in Computer and Information Science* (Springer, 2012), pp. 192–199. doi:10.1007/978-3-642-32427-7_26

21. R. Velmani, B. Kaarthick, An energy efficient data gathering in dense mobile wireless sensor networks. ISRN Sens. Netw. (2014). doi:10.1155/2014/518268
22. R. Velmani, B. Kaarthick, An efficient cluster-tree based data collection scheme for large mobile wireless sensor networks. IEEE Sens. J. **15**(4) (2015)
23. N. Bagga, S. Jain, T.R. Sahoo, A cluster-tree based data dissemination routing protocol, in *Eleventh International Conference on Communication Networks, ICCN 2015, ICISP 2015, ICDMW 2015*, 21–23 Aug 2015, Bangalore, India, vol. 54 (Procedia Computer Science, Elsevier, 2015), p. 7
24. S. Sharma, *On Energy Efficient Routing Protocols for Wireless Sensor Networks*. Ph D Thesis. Department of Computer Science and Engineering National Institute of Technology, India (2016), http://ethesis.nitrkl.ac.in/6962/1/2016_Suraj_Phd_509CS607.pdf. Accessed 11 Nov 2016
25. A.B.M.A. Al Islam, C.S. Hyder, H. Kabir, M. Naznin, Finding the optimal percentage of cluster heads from a new and complete mathematical model on LEACH. Wirel. Sens. Netw. **2**, 129 (2010). doi:10.4236/wsn.2010.22018
26. W.R. Heinzelman, A. Chandrakasan, H. Balakrishnan, Energy-efficient communication protocol for wireless microsensor networks, in *Paper presented at the Hawaii International Conference on System Sciences*, Maui, Hawaii, 4–7 Jan 2000 (2000)
27. M. Gauger, *Coordinate assignment, in Integration of Wireless Sensor Networks in Pervasive Computing Scenarios* (Logos Verlag Berlin GmbH, 2010)
28. Z. Li, R. Li, Y. Wei, T. Pei, Survey of localization techniques in wireless sensor networks. Inf. Technol. J. **9** (8) (2010)
29. A. Bensky, Received signal strength, time of arrival and time difference of arrival, in *Wireless Positioning Technologies and Applications* (Artech House Publishers, 2008)
30. S.P. Singh, S.C. Sharma, Range free localization techniques in wireless sensor networks: a review, in *3rd International Conference on Recent Trends in Computing 2015 (ICRTC-2015)*, vol. 57 (Procedia Computer Science, 2015). doi:10.1016/j.procs.2015.07.357. p 7
31. J. Bardwell, *The Truth About 802.11 Signal and Noise Metrics* (2004), http://www.n-cg.net/ncgpdf/WiFi_SignalValues.pdf. Accessed 1 Aug 2016
32. A. Boukerche, H.A.B.F. Oliveira, E.F. Nakamura, A.A.F. Loureiro, Localization systems for wireless sensor networks, in *Algorithms and Protocols for Wireless Sensor Networks*, Hoboken, NJ, USA (Wiley, 2007). doi:10.1002/9780470396360
33. M. Arbabi, M., Reza, Abrishambaf, S. Uysal, Localization in Wireless Sensor Networks Based on Sugeno Fuzzy Logic. MS Thesis (2013), http://i-rep.emu.edu.tr:8080/xmlui/bitstream/handle/11129/173/Monfared.pdf?sequence=1. Accessed 1 Aug 2016
34. D.F. Larios, J. Barbancho, F.J. Molina, C. León, LIS: Localization based on an intelligent distributed fuzzy system applied to a WSN. Ad Hoc Netw. **10** (2012) (Elsevier)
35. A. Kumar, V. Kumar, Fuzzy logic based improved range free localization for wireless sensor networks. Int. Sci. Index, Electr. Comput. Eng. **7**(5) (2013)
36. Veris White Paper VWP18, *Veris Aerospond Wireless Sensors: Received Signal Strength Indicator (RSSI)*. Veris Industries (2013), http://www.veris.com/docs/whitePaper/vwp17_WiFi%20vs%20Zigbee_RevA.pdf. Accessed 4 Aug 2016

Analyzing Slavic Textual Sentiment Using Deep Convolutional Neural Networks

Leo Mršić, Robert Kopal and Goran Klepac

Abstract In this paper we deploy a deep architecture for convolutional neural networks for understanding of Croatian. We follow the same approach as in [34], and we use deep learning on character inputs on a sentiment analysis dataset in Croatian. Although we have archived considerable results (without any complex parsing or background knowledge), the result was inferior to that reported in the abovementioned paper. As Croatian is one of the low-resource languages, there are considerable links between using such an approach (that maximizes the role of data) and sustainability. The main objective of this chapter is to give a clear understanding of the position of low-resource languages and propose a direction for sustainable development of language technologies illustrated using convolutional neural networks for textual sentiment analysis. Impact of this research to scientific but also business community is significant due to fact that every method with acceptable ratio of simplicity and effectiveness can be included inside more complex logic environment focusing on specific language or appliance in specific area. Since language structure is generally not easy to manage, there is constant need for improvement of tools used to score text data especially while majority of unstructured data analysis tools often transfer various data to text and create further analysis paths form there.

Keywords Deep convolutional neural networks · Sentiment analysis · Slavic languages · Sustainable techniques

L. Mršić (✉) · R. Kopal · G. Klepac
University College Algebra, Zagreb, Croatia
e-mail: leo.mrsic@racunarstvo.hr

R. Kopal
e-mail: robert.kopal@racunarstvo.hr

G. Klepac
e-mail: goran.klepac@racunarstvo.hr

© Springer International Publishing AG 2017
A.K. Sangaiah et al. (eds.), *Intelligent Decision Support Systems
for Sustainable Computing*, Studies in Computational Intelligence 705,
DOI 10.1007/978-3-319-53153-3_11

1 Motivation

Following fast development of different contextualization tools and services and its appliance in real life, effort increasing trend in text analysis can be recognized almost daily. To be able to manage direct commands and/or keywords is especially challenging considering special language structure for languages different than most popular ones. Slavic languages structure is quite complex and not easy to follow for users not used to so many variations and word bases. Therefore, sentiment analysis became more complex to control while at the same time more challenging to manage, train and use. Objectives for this research is to explain complexity and characteristics of Slavic languages important for analysis and to challenge methods which provide best results from researcher's experience so far. Research focus is to find suitable level of analysis management reaching feasible ability to use results in related tools and services. Motivation for such research is in great potential in understanding of Slavic languages which are out of general focus while at the same time user needs are the same for most popular languages. To be able to manage different methods on Slavic languages (textual sentiment as example in this research) will be great opportunity for integration such services in common products.

2 Introduction

Natural language understanding is one of the long term goals for AI set forth at its very beginning at the Dartmouth conference. Natural language understanding is often understood in its symbolic form as for example in [6], but we are interested in a connectionist approach to natural language understanding, in the sense that we want to use only neural models. We also want to emphasize, that although much work has been done on the topic of extracting structural information (such as Named Entity Recognition and Part of Speech tagging), in this paper we are interested solely in sentiment analysis. For the dataset, we have chosen the dataset of [7], which is the largest sentiment analysis dataset available in Croatian, and there abovementioned paper also gives a benchmark. Our results give a significant improvement but fall short of the results reported in [34].

One notable algorithm which we will not explore, but which deserves to be mentioned is word2vec [16]. We want to emphasize how our approach differs from the usual approaches using word2vec. The traditional approaches with word2vec are usually coupled with intensive feature engineering, such as [9] or [16]. Following [34] we use temporal ConvNets [17], for input we pass individual characters and the output is a Boolean sentiment value, where 0 denotes negative sentiment and 1 denotes all other text (as was done in cite [7]). As the authors of [34] point out, the inspiration of this approach is machine vision, where individual pixel values are used as input for e.g. recognizing cats on pictures.[1] The reason why we do not use

---

[1]Cf. http://www.pyimagesearch.com/2016/06/20/detecting-cats-in-images-with-opencv/.

Word2vec is because the approach from [34] we follow is less intensive in computational terms and needs only trivial parsing to characters, which offers a sustainable approach.

2.1 Slavic Languages Analysis Characteristics

A low resource language is a language which does not have large available datasets and benchmarks for most natural language processing[2] tasks. A case can be made that a low resource language is a paradigm for sustainable computing. In the modern age low resource language users have roughly the same demand as resource rich language users, and unlike resource-rich languages, low-resource language technologies cannot afford to be "hacky". In a sense, everything must be done right for the first time since there is no possibility for incremental and slow development. As such, low-resource language approaches can be a beacon which shows the way things should be done all languages.

Slavic languages form a large language group spoken in Central and Eastern Europe, and make a language group that is more homogenous than the other language groups. Up until the 10th century, all Slavic people spoke a single language, commonly referred to as Common Slavic. Today, there are three major language groups (Western, Eastern and South) totaling about 20 languages, 10 of which have an official status in one of the European countries. The number of speakers of Slavic Languages is estimated at 315 million worldwide [19]. The eastern group comprises of Russian, Ukrainian, Belarusian, the western group comprises of Polish, Czech and Slovak, while the southern group consists of Slovene, Serbian, Croatian, Macedonian and Bulgarian. There is speculation about the possible existence of the North Slavic branch, but these discussions are beyond the scope of this paper.

Slavic languages are spread across a multicultural area, encompassing all the major European religions, and influences from other cultures. Slavic languages are traditionally low resource languages. The reason why remains elusive, since the Soviet Union as the most populous Slavic country in the past century could have (in principle) initiated a technological race with the United States in language technology. Machine translation was a major driving force in the US during the Cold War, which in turn funded much of the research in computational intelligence over the last 50 years. The Soviet Union had no computational linguistics program and no similar ambition. Slavic languages share a different structure than English. They have classes and declension, they have a complex morphology annotating tenses with suffixes rather than by composition of auxiliary verbs and they have a number of formation methods like prefixation which lend themselves well to pattern recognition techniques.

---

[2]In the following we shall use the abbreviation NLP to stand for "natural language processing".

The problem with developing standard natural language processing tools for Slavic languages is in par the great degree of fragmentation in practically all Slavic states except Russia, and this poses a problem since efforts are not easily combined and coordinated. The approach that takes in account sustainability as one of the major challenges could in practice serve as a guideline for all other Slavic languages, and motivate approaches requiring only a limited amount of labor and data preprocessing. The main paradox regarding Slavic languages is that the adoption of foreign words in regular usage is higher in the languages which maintain the highest degree of similarity to Common Slavonic, such as for example the Chakavian dialect of Croatian, while it is the lowest with the languages that were most changed over time, such as for example Russian. We could only speculate on the reasons behind this, but it is an important fact which poses serious problems for a unified resource expansion by using machine translation technologies that would import corpora from resource rich languages such as English.

This might seem strange at first, but many natural language processing tasks are quite insensitive under machine translation. For example, if one could map with a bijective mapping English and Portuguese words in a given sentence, even though the translation might be horribly artificial, for most machine-learning tasks focusing on word-level features, this might work, since it is (as we have assumed) a one to one mapping. One great example where this approach must work is for named entity recognition, which is the task of automatically recognize the words that correspond to entities. If we have a starting string AAA BBB CCC in a source language, and a machine translation system mapping AAA to XXX, BBB to ZZZ and CCC to YYY, where XXX, ZZZ and YYY are strings in the target language, then if BBB refers to a named entity, so does ZZZ. The reason why this approach works well for Named Entity Recognition is because the entities are external in a robust sense: naming a location is very often done with a single word or sequence of words and almost never by modifying other words. Sentiment analysis, on the other hand suffers greatly from the problem of absorption in other components.

This is especially true of Slavic languages [11], which are often rich in sarcasm and irony, and most often so by modifying the surrounding words and the general context in which the irony subsists. Irony can often be addressed in terms of polarity [2], but sarcasm is harder to detect and eludes most techniques except deep learning. The point to note is that irony means saying the inverse of what one feels and this is easily detectable by neural language models, whilst sarcasm is absorbed in the choice of words and word combinations both of which are virtually indistinguishable at text level since it requires genuine understanding and, more often than not, reasoning and knowledge. This means that for building a sarcasm detection system one would need to use a complex cognitive system based on symbolic methods capable of reasoning and inference, which is currently out of reach for subsymbolic systems of deep learning.

2.2 Data and Privacy

In this section we explore two arguments from [10] in the context of the vindication of rights of machines, and develop a third different argument based on a technological singularity in the context of artificial neural networks and their learning mechanism. We explore how the learning process alters the existing algorithm and how this voids the standard perception of a moral agent since a trained neural network becomes demonstrably different than untrained algorithms, and more similar to humans and animals than to regular software when considering self-improvement as an essential attribute of living beings.

Moral questions regarding machine learning and big data privacy is a new theme. Sustainability in a broad sense can only be archived if a common framework is defined to make sense of privacy, data ownership and algorithm ownership issues. The question of algorithm rights is explored in [10]. Vindication-of-rights arguments, according to [10] always share the same structure. First new arguments are enacted which augment the criteria of membership in the class of moral subjects, and then it is shown that according to that criteria some new entities might claim membership. This is an extension of the general "liberation movement" of Singer [29].

Gunkel takes an atypical view of the vindication of rights and points out that a criterion of membership is not clear. A second problem which Gunkel finds is that a view solely based on the moral agent and not taking in account the "moral sufferer" is one-sided. This argument comes from the animal rights vindication movement [29].

If one was to extend these arguments in the context of computational intelligence (algorithms, big data, privacy and ownership) one would follow Singer in assessing that biological configuration cannot be the sole basis for moral evaluation and ownership claims. If we take that the creator has exclusive rights over a given algorithm, but not over an essentially different algorithm, we have to redefine our usage and practices to archive sustainable results. An interesting point about sustainability and computational intelligence is that computational intelligence systems, unlike regular software systems get better with age. Sustainability then becomes of paramount importance, since the time gone in training them and the data filtered cannot be easily replaced.

3 Textual Sentiment Analysis: Methods

Computational intelligence started as a minor field at the Dartmouth conference in 1956, and the idea of Computational Intelligence was primarily driven by trying to make the computers archive some of the tasks that humans considered to be intelligent. This was clearly stated in its manifest:

> The study is to proceed on the basis of the conjecture that every aspect of learning or any other feature of intelligence can in principle be so precisely described that a machine can

be made to simulate it. An attempt will be made to find how to make machines use language, form abstractions and concepts, solve kinds of problems now reserved for humans, and improve themselves. We think that a significant advance can be made in one or more of these problems if a carefully selected group of scientists work on it together for a summer.[3]

The proposal proceeds to list all the fields that would later become the backbone of Artificial Intelligence, and Computational Intelligence:

[...] 2. How Can a Computer be Programmed to Use a Language: It may be speculated that a large part of human thought consists of manipulating words according to rules of reasoning and rules of conjecture. From this point of view, forming a generalization consists of admitting a new word and some rules whereby sentences containing it imply and are implied by others. This idea has never been very precisely formulated nor have examples been worked out. 3. Neuron Nets: How can a set of (hypothetical) neurons be arranged so as to form concepts. Considerable theoretical and experimental work has been done on this problem by Uttley, Rashevsky and his group, Farley and Clark, Pitts and McCulloch, Minsky, Rochester and Holland, and others. Partial results have been obtained but the problem needs more theoretical work. [...] 5. Self-Improvement: Probably a truly intelligent machine will carry out activities which may best be described as self-improvement. Some schemes for doing this have been proposed and are worth further study. It seems likely that this question can be studied abstractly as well. 6. Abstractions: A number of types of "abstraction" can be distinctly defined and several others less distinctly. A direct attempt to classify these and to describe machine methods of forming abstractions from sensory and other data would seem worthwhile. 7. Randomness and Creativity: A fairly attractive and yet clearly incomplete conjecture is that the difference between creative thinking and unimaginative competent thinking lies in the injection of a some randomness. The randomness must be guided by intuition to be efficient. In other words, the educated guess or the hunch include controlled randomness in otherwise orderly thinking.

In its beginnings, computational intelligence was primarily "applied philosophy" and only in second place engineering. This is well illustrated by the quote of Herbert Simon and Alan Newell (the only people at the conference that actually had a working AI algorithm):

[We] invented a computer program capable of thinking non-numerically, and thereby solved the venerable mind-body problem, explaining how system composed of a matter can have the properties of mind.

One subdiscipline of Computational Intelligence that has gained a lot of traction of the last decade, but was present from the very beginnings are artificial neural networks. Neural networks are typically supervised learning algorithms [12] thattake data together with labels and produce labels for new unseen data.

The beginning of neural networks can be found in [21], but the first important improvement came from Hebbian learning [13]. The basic elements of neural networks are perceptrons [24]. The main idea was that a single perceptron consists of numerous inputs $(a_0, a_1, \ldots, a_n)$ which have a value between 0 and 1. Each of them has its own weight $(w_0, w_1, \ldots, w_n)$ which can be understood as the degree of importance of a given connection. Once all the inputs and weights are multiplied, they are added up and some value is obtained which is then passed to a perceptron which has

---

[3] See http://www-formal.stanford.edu/jmc/history/dartmouth/dartmouth.html.

a threshold (e.g. 10). If the threshold is met, the perceptron activates and sends 1 forward, else it sends 0.

Minsky and Papert [22] first saw the limitations of these simple models and proved that some functions cannot be calculated in this manner. These problems were solved by connecting perceptrons in layers. This is archived by forming layers of neurons (modified perceptrons) in a shallow multilayer perceptron having an input layer, a hidden layer and an output layer. The 1980s showed a renewed interest in neural networks, and some of the results obtained like the Boltzmann Machines [3] are still in use today. But the most notable improvement made was the discovery of backpropagation. Although originally discovered by Werbos [31] it remained unnoticed by the scientific community. It was rediscovered by David Parker in 1981 but published only later in [23], and independently by LeCun [18] published in French and again discovered by Hinton in [14]. During the 1990s there was a decline in interest in neural networks in favor of other machine learning approaches, most notably the Support Vector Machines. In the late 2000s neural networks came in favor again as huge amounts of data were being put to use, and several of the top IT companies adopted neural networks and began developing custom hardware.

3.1 Basic Linear, Binary, ReLU, Tanh and Sigmoid Model

The simplest form of a neuron is the linear neuron described below:

$$y = \sum_i x_i w_i + b$$

This simple model has many shortcomings, one of the major one is that it has no decision procedure, i.e. all one gets as the output is the weighted sum of the inputs and a bias term. To allow it to make decisions and to be a classifier, it must be able to produce a classification. To introduce a decision we define a binary threshold neuron:

$$z = \sum_i x_i w_i + b$$

And:

$$y = 1, \text{ if } z \geq \theta$$

And 0 otherwise. A simple modification of this is the rectified linear unit (ReLU), where $y = z$, if $z > 0$ and 0 otherwise. Sigmoids are passed through a logistic function:

$$z = \sum_i x_i w_i + b$$

And:

$$\frac{1}{1 + e^{-z}}$$

The tanh function is similar to the sigmoid, but defined as:

$$y = \frac{e^z - e^{-z}}{e^z + e^{-z}}$$

3.2 Basic Neural Network Deployment Setting

One of the most basic datasets for neural network testing and proof-of-concept is the so called MNIST dataset which consists of handwritten digits, with the label which contains the correct number. We give a brief overview of the neural network functionality exemplified on a MNIST-like task. Images are formatted to a 20 by 20 size, and grayscaled. For a 20 by 20 image, 400 input neurons are needed, and it is assumed that pixels are enumerated left to right, up to down. This means that every pixel will have a single value in the 0–1 interval, where 0 is white and 1 is black. The hidden layer can contain as much hidden neurons as possible, but keeping in mind the more neurons it has the more time will it take to train. Two output neurons are needed, and their values are to be interpreted as confidence scores for YES and NO, not as probabilities.

The most important feature of neural networks is its training over multiple layers, which was implemented in 1975 [31]. To train the neural network to recognize digits, a dataset with labelled images is needed, and it has to have plenty of examples. 50000 is a good rule of thumb. For each image, ne network tries to classify the image, getting the result e.g. YES = 0.34 and NO = 0.87. The cost is calculated and the error is backpropagated with partial derivatives to each weight. Technically speaking, what was the threshold in perceptrons is now called a bias, and all the weights and activations remain the same, so the chain rule for derivatives may be used.

Each passing of the backpropagation algorithm is called an epoch of training and it is customary to train for around 50 epochs. One of the main questions is whether we need more hidden layer neurons or more epochs to make things better. Newer results suggest that neural networks with multiple hidden layers consistently outperform the ones with a single hidden layer. An observation can be made that during learning a neural network changes, and it does so in an essential fashion. The same basic network can be trained to be good at radically different tasks, and it can be trained by exactly the same methods (only the input data differs). This opens multiple issue over claiming ownership over neural networks, and also about the usefulness of using them for an extended period of time.

3.3 Backpropagation

Backpropagation is the learning procedure of choice in artificial neural networks. We start by defining the residual error as the error between expected results and actual results. The function used is the squared error function:

$$E = \frac{1}{2} \sum_{n \in Train} (t^n - y^n)^2$$

Differentiating with respect to one of the weights will give:

$$\frac{\partial E}{\partial w_i} = \frac{1}{2} \sum_n \frac{\partial y^n}{\partial w_i} \frac{dE^n}{dy^n}$$

where:

$$\frac{\partial y^n}{\partial w_i} = x_i^n$$

And:

$$\frac{dE^n}{dy^n} = 2(t^n - y^n)$$

Then the batch delta rule changes the weights in proportion to their error derivatives summed over all training cases:

$$\Delta w_i = -\eta \frac{\partial E}{\partial w_i} = \sum_n \eta x_i^n (t^n - y^n)$$

For logistic sigmoid neurons the procedure is a bit more complicated. We start by noting that:

$$z = b + \sum_i x_i w_i$$

And that:

$$\frac{\partial z}{\partial w_i} = x_i$$

And:

$$\frac{\partial z}{\partial x_i} = w_i$$

Then for the logistic function:

$$\frac{1}{1 + e^{-z}}$$

It holds that:

$$\frac{dy}{dz} = y(1 - y)$$

This clearly holds since:

$$y = \frac{1}{1 + e^{-z}} = (1 + e^{-z})^{-1}$$

Note that:

$$\frac{dy}{dz} = \frac{-1 \cdots (-e^{-z})}{(1 + e^{-z})^2} = (\frac{1}{1 + e^{-z}})(\frac{e^{-z}}{1 + e^{-z}}) = y(1 - y)$$

This is clear since:

$$\frac{e^{-z}}{1 + e^{-z}} = \frac{(1 + e^{-z}) - 1}{1 + e^{-z}} = \frac{1 + e^{-z}}{1 + e^{-z}} - \frac{e^{-z}}{1 + e^{-z}} = 1 - y$$

When the derivative is calculated, the output with respect to the logit and the derivative with respect to the weight, it is possible to calculate the derivative of the output with respect to the derivative of the weight. To do this, the chain rule is needed:

$$\frac{\partial y}{\partial w_i} = \frac{\partial z}{\partial w_i}\frac{dy}{dz} = x_i \cdot y(1 - y)$$

And with this the learning rule for the logistic neuron is obtained:

$$\frac{\partial E}{\partial w_i} = \sum_n \frac{\partial y^n}{\partial w_i}\frac{\partial E}{\partial y^n} = -\sum_n x_i^n y^n (1 - y^n)(t^n - y^n)$$

Which looks remarkably similar to the delta rule, and one can see that the extra term $y^n(1 - y^n)$ is actually the slope of the logistic. This can be further modified to encompass the idea of reinforcement learning to be able to learn not just from the labels but from the arrival of data itself.

Reinforcement learning is the type of learning when instead of the error function there is a reward function which has to be maximized. The reward function is usually sensitive to a severe time lag, but if online learning is used to build a distributed representation of the data as data comes in, it can prove to be quite simple to implement.

3.4 Convolutional Neural Networks

Neural networks are a type of machine learning algorithm which uses weights and biases to classify data. The weights and biases are updated via backpropagation of classification errors, usually using gradient descent (although several other options like contrastive divergence are also being used). For years neural networks where made of one input layer one hidden layer and one output layer, since there was a problem when training multiple hidden layers with gradient descent. In recent years, this problem has been solved by using forms of unsupervised pre-training like for example autoencoders. The way autoencoders help is by modifying the distributed representations of concepts in their layers making them more "digestible". The question of why exactly this happens remains open.

Convolutional neural networks are a type of deep architecture which does not depend on unsupervised pretraining. This is because convolutional networks work in a special manner. The main part of convolutional networks are convolutional modules. Convolutional modules are modules made up of a small neural network (n*m in input neuron size, and a single output neuron) which goes across data and learns to detect the same pieces of information throughout the data. The outputs constitute a feature map. The convolutional operator can scan the input data N times (each time with new randomly initiated weights), and this will produce N feature maps.

The convolutional operator as stated contains n*m weights for the receptive field neurons and a single bias (for the output neuron). The output feature map is usually passed to a pooling layer. The pooling layer takes a receptive field of its own and chooses which value to pass from the receptive field. This is most often simply the maximal value from the receptive field, and such a layer is called a max pooling layer. Convolutional neural networks archived great success in machine vision, but they are also implemented in other areas. We explore one of the most novel application, using neural networks for natural language processing of low resource languages, where stemmers and parsers are not available or are rudimentary. Our approach has the benefit of avoiding this problems by analyzing the text character by character.

3.5 Sentiment Analysis

The idea of tracking sentiment on a large scale first came from the financial markets as reported in [27]. The idea flourished with the advent of e-commerce as a way of monetize consumer reviews. The idea was quickly reintegrated in mainstream natural language processing. Sentiment analysis can be either machine learning based, rule-based or hybrid. Most state-of-the-art deployments today are hybrid, although more often than not, deployed systems are marketed as fully machine learning based. Sentiment analysis systems take as input a text and then classify them as positive or negative (or neutral), or provide a scale (e.g. -5 to 5). The maximal reported agreement of humans on sentiment analysis [32] is around 82%, which means that above that is super-human performance [26], which in the context of natural language processing does not make much sense. Today there are numerous approaches to sentiment analysis hand-taylored for low resource languages like for example [5] or [1], but none of these approaches take in account the need for sustainability (by deploying a minimalist system), and we offer our own approach.

3.6 Problem with Deep Structures

One of the major problems with deep architectures is the problem of vanishing and more rarely exploding gradients. Vanishing gradients occur when the iteration of differentiation during backpropagation decreases the gradient across layers and it comes to nearly zero at the end. One popular method of combating this is using unsupervised pretraining of the hidden layers, and this is the best way to do it. Other strategies include Adding the logits across layers so that they are perceived by the upper layers. This is a good approach but it is still in its infancies. The problem of exploding gradient happens when the learning examples are very similar and then the gradient explodes, due to the fact that moving in the direction of the gradient actually increases the error, and by iteration the gradient becomes larger and larger. The most modest way to combat this is to use a convolutional neural network which can lead to a simpler but deep architecture due to the fact that only a part of the model is being trained and the number of neurons are limited to the receptive field neurons, which is always less than the usual number of neurons in a standard neural network. Attention units can also be used to enhance performance, but they are beyond the scope of this paper.

4 Textual Sentiment Analysis in Slavic Languages: The Model

The model we used was as the one in [34], a simple temporal convolutional module which computes a 1D convolution, with max pooling. Formally, let $g(x)$ be a discrete input function mapping an interval $[1, i]$ to the whole set of reals, and $f(x)$ be a discrete kernel function mapping $[1, j]$ to the set of reals. Let d be the convolution stride, the convolution $h(x)$ which maps $[1, ((i - j + 1)/d)]$ is defined as:

$$h(y) = \sum_i f(x) \cdot g(y) \cdot d - x + c$$

where $c = j - d + 1$ is the offset constant. The module is parameterized by kernel functions fmn which are called colloquially weights on a set of inputs $gm(x)$ and outputs $hn(y)$ (each of these is called an input or output frame). We departed from [34] in that we explored three nonlinearities, namely ReLUs, hyperbolic tangent functions and sigmoids rather than just ReLUs. We have used stochastic gradient descent as the authors did in [34]. As in [34], character quantization was made using ASCII characters only (the Croatian specific characters were recast in ASCII). Both models are 9 layers deep, 6 convolutional and 3 fully connected layers at the end, and they differ in the number of hidden neurons. Max pooling was used as well as dropout between the fully connected layers (with a 0.5 probability). We have normalized the input with the mean 0 and STD 0.05. No data augmentation was possible, and we skipped on the idea of using WordNet as there was not enough resources for Croatian to contribute significantly.

4.1 Model Results and Comparison with Dataset Benchmarks

The English classificator in [34] was trained on the Amazon Review dataset spanning over 18 years with 34 million reviews on more than 2 million products. Only texts in size between 100 and 1014 were chosen, and 3 star reviews were left out. The 1 and 2 star reviews were considered negative and 4 and 5 are considered positive. Out of the remaining reviews, 1275000 texts with one and five stars and 725000 of two and four star reviews were selected at random. The size of the training set was 3.6 million, while for testing 400000 test samples were used. They obtained the following results (in terms of accuracy) (Tables 1 and 2).

For Croatian we have used the Croatian Telco sentiment analysis dataset presented in [7], which is composed of negative and other texts. The preliminary benchmarks for the datasets are accuracy 68.14%, precision 95.23% and recall 36.36%.

These results show that across the test, Sigmoid performed slightly better than the other contenders, and large networks were better over small ones. The average training time of the large network was around 5 days per epoch, while for the smaller ones around 2 days, which puts in perspective the small gains. Also it must be noted that ReLU's trained on average 25% quicker per epoch. In total, we do not have any clear winner, and the choice of the type of network is dependent on other factors. In the context of energy consumption, Small ConvNets with ReLU's are the most sensible choice. In terms of test scores Sigmoids outperformed the others in both small and large networks. It should be repeated that according to [32] the maximal unanimous sentiment classification by humans is around 82%.

Table 1 Accuracy for different models tested

Model	Thesaurus	Training accuracy (%)	Test accuracy (%)
Large ConvNet	No	62.96	58.69
Large ConvNet	Yes	68.90	59.55
Small ConvNet	No	69.24	59.47
Small ConvNet	Yes	62.11	59.57
Bag of words	None	54.45	54.17
Word2vec	None	36.56	36.50

Table 2 Our model accuracy, precision and recall

Model	Activation	Training accuracy (%)	Test accuracy (%)	Precision (%)	Recall (%)
Benchmark	None	Not reported	68.14	95.23	36.36
Large ConvNet	ReLU	58.45	55.67	88.45	39.89
Large ConvNet	Tanh	**59.51**	57.11	84.67	39.00
Large ConvNet	Sigmoid	58.98	**58.01**	**89.09**	**41.47**
Small ConvNet	ReLU	57.33	53.56	79.89	36.78
Small ConvNet	Tanh	55.56	54.07	83.44	35.33
Small ConvNet	Sigmoid	**58.09**	**57.90**	85.02	**37.21**

4.2 Model Sustainability

The current definition of sustainability is the intersection between economic development, social development and environmental protection. The intersection between the first two is equitable development, the intersection between social development and environmental protection is bearable development and the intersection between environmental protection and economic development is viable development [20].

These are the definitions set forth by the 2005 World Summit on Social Development and they make clear that sustainable development must be both bearable, in terms of social and environmental issues, equitable, in terms of social and economic issues and a viable interaction of economic goals and environmental considerations.

What is most interesting for our research is the fourth ingredient that is sometimes added, culture. As language is considered to be an integral part of a nation's culture, language processing, especially sustainable language processing has to happen at the intersection of environmental issues such as power and storage, economic issues such as novelty and impact, and social issues such as usability. Only by meeting these four requirements, can sustainability happen, and for language related technologies, the main one is cultural acceptance, and perhaps the least important (but not unimportant) is the economic impact, since economic impact can seldom be the driving cause of high profile research over the long term.

4.3 Data Sustainability and Big Data

The question of sustainability is closely connected to the question about data in another matter as well. Can we use the data we already have to optimize losses (power, storage, etc.), and if not can we offset the impact of collecting more data with future savings in terms of resources. An important and relevant question is what is data, and more importantly what is big data. There is no single definition of big data on the market, but the received wisdom claims that big data is data of the size exceeding any concrete storage, and exceeding the possibility of browsing data by any particular user [8]. When addressing any particular data point it seems obvious to whom does the data belong, and above all who has the right to claim a profit derived from the data. In big data, these problems become more complex due to their sheer intractability. It is also possible to ask the question whether the focus in big data is on "big" or "data", and according to our answer, a certain approach will be preferable [8]. We claim that in the context of sustainability, the emphasis should be on "big" and on avoiding data duplication and augmentation. A number of researchers tried to evaluate the value of the information around us and the speed at which it grows. The results varied but converged to a single point: Big data is the source of new economic value and innovation [28]. In this context, using a sustainable approach like the one we envisioned for structuring unstructured data becomes of paramount importance, especially considering that our approach is saving labor and time by not needing data augmentation or preprocessing. Organizations will have to care for data access which are relevant to their goals and for which big data technologies will be deployed. They can be obtained from any source including structured data from relational databases or unstructured data from textual files, or even from online sources. Whatever their sources might be, organizations should enable permanent access to this data so that the data at hand can be used continuously [15]. A different issue of cultural sustainability also arises. The big data content that gets replicated and has the biggest impact on sustainable goals is actually the culturally least significant portion, and it has the potential to overshadow culturally significant parts. The culturally significant parts of big data are often quality textual material, something akin (but not identical to) literature. By providing tools which can recommend not just the "relevant" popular text content, but also the relevant and culturally significant texts, these sustainability issues will also be addressed.

4.4 Privacy Issues

Privacy issues are a central concern today, and many fields have undertaken extensive measures (both in terms of policy and implementation) to guarantee data privacy, from the financial industry [4], requirements engineering [33] and most relevant to our own research, internet technologies [30]. A very interesting idea which connects sustainability and privacy is reported in [25], where measures are undertaken to enable stricter legal regulation of data privacy by specifying what deployment systems need to have incorporated in them.

The big data approach differs from the question of structuring data from the viewpoint of privacy and data protection. There are two fundamental differences which make a difference: (1) big data analysis is usually not conducted for the same reason the data is collected and (2) the amount of data used for processing big data greatly exceeds the abilities of traditional databases (with tested safeguards). The primary goal of big data is deriving new insights and predicting the behavior or pattern analysis on data from multiple sources. The differences in regulation determine two key questions (1) the possibility of accessing data if access has been granted and (2) the question for what purpose would a given piece of data be used. The process of anonymization is impossible in the context of big data, since it is possible to uniquely re-identify each individual.

5 Conclusion

The approach in [34] is particularly well suited for low resource languages in terms of sustainability and sustainable development. The main reason for this is the fact that this approach is a simple machine learning approach which requires little dedication in terms of data preparation and manual feature engineering. Also, there is the option to choose a simpler, but less effective approach and a smaller network. We have shown that this is indeed possible and that the loss in performance is not that spectacular. For a majority of tasks it is possible to use the more computationally efficient parameters. The same approach used here can be used for any low resource language. A note must be made that sustainability has unintended additional meaning which our research does not support. Sustainability can be interpreted as accepting the status quo, which is something we do not do nor promote. The true idea of sustainability embraces development and high profile ideas to lessen the negative impacts of human activity, and in this sense we feel our research has also made a contribution. The question of the interaction between computational intelligence and sustainability will remain tense for years to come. As computational intelligence advances, pressures will be put on al resources needed, be it human resources, power, computer hardware and even secondary pressure to educational resources such as paper and consumables. It is clear that the process will also put a great strain on data resources, as most of the work done in computational intelligence today is data-driven, and there are questions related to sustainability which have to be answered before the stress becomes so high as to invalidate certain less invasive methods. The data collected is enough to fuel the development of the data driven technologies of the high-resource languages, but there is great inefficiency there in terms of wasteful data-driven research which consumes a lot of hardware, manpower and electrical power, which could better be hand-coded. It is understandable that hand-coding is undesirable from the view of applications, modern trends and financial cost, but from the point of view of sustainability, it is a much better approach. What we have offered is to combine the two approaches, by deploying applications that are data-driven, but at the same time more sensible in terms of sustainability when it comes to low resource languages, keeping in mind that technologies which are constructed and intended to remain low resource are much more desirable than scaling up the resource collection and storage, and consequently transforming a low resource language into a high resource language to be able to use high resource tools available for English. This approach could still be made feasible by employing massive scale reinforcement learning to maximize the information gain as the resources come in sequentially. There has been much research going on unsupervised learning, which is very promising since it cuts on the manpower needed for learning, but these techniques are not nearly as good as classical supervised learning. We proposed a system which is supervised learning, and which can be quickly adapted to use simple text, circumventing the need for more intricate features. This saves a lot of manpower, and we hope that future research will increase the performance of this approach and also investigate similar sustainable approaches for low-resource languages.

References

1. A. Abbasi, H. Chen, A. Salem, sentiment analysis in multiple languages: feature selection for opinion classification in web forums. ACM Trans. Inf. Syst. (TOIS) **26**(3) (2008)
2. M. Abhijit, K. Diptesh, N. Seema N., Kuntal, B. Pushpak, Leveraging cognitive features for sentiment analysis, in *Proceedings of The 20th SIGNLL Conference on Computational Natural Language Learning* (2016)
3. D.H. Ackley, G.E. Hinton, T.J. Sejnowski, A learning algorithm for boltzmann machines. Cogn. Sci. **9**, 147–169 (1985)
4. A. Anton, J. Earp, Q. He, D. Bolchini Stufflebeam, C. Jensen, Financial privacy policies and the need for standardization. Secur. Priv. IEEE **2**(2), 36–45 (2004)
5. C. Banea, R. Mihalcea, J. Wiebe, S. Hassan, Multilingual subjectivity analysis using machine translation, in *Proceedings of the Conference on Empirical Methods in Natural Language Processing (EMNLP-2008)* (2008)
6. P. Blackburn, J. Bos, *Representation and Inference for Natural Language: A First Course in Computational Semantics* (Center for the Study of Language and Information, Amsterdam, 2005)
7. B. Dropuljic, S. Skansi, R. Kopal, Analyzing affective elements using acoustic and linguistic features, in *Proceedings of the Central European Conference on Information and Intelligent Systems (CECIIS 2016)* (2016), pp. 201–206
8. B. Franks, *Timing the Big Data Tidal Wave: Finding Opportunities in Huge Dana Streams with Advanced Analytics* (Wiley and SAS Business Series, New York, 2012)
9. A. Frome, G. Corrado, J. Shlens, J. Dean, T. Mikolov, Devise: A deep visual-semantic embedding model, in *Advances in Neural Information Processing Systems* (2013), pp. 2121–2129
10. D.J. Gunkel, A vindication of the rights of machines. Philos. Technol. **27**(1), 113–132 (2014)
11. J. Haiman, *Talk Is Cheap: Sarcasm, Alienation, and the Evolution of Language* (Oxford University Press, Oxford, 1998)
12. M. Hassoun, *Fundamentals of Artificial Neural Networks* (MIT Press, Cambridge, 2003)
13. D. Hebb, *The Organization of Behavior* (Wiley, New York, 1949)
14. G.E. Hinton, Learning distributed representations of concepts, in *Proceedings of the Ninth Conference of the Cognitive Science Society*, Amherst, MA (1986), pp. 1–12
15. J.R. Kalyvas, M.R. Overly, *Big Data: A Business and Legal Guide* (Taylor and Francis Group, New York, 2015)
16. Q. Le, T. Mikolov, *Distributed Representations of Sentences* (2014). arXiv:1405.4053
17. Y. LeCun, L. Bottou, Y. Bengio, P. Haffner, Gradient-based learning applied to document recognition. Proc. IEEE **86**(11), 2278–2324 (1998)
18. Y. LeCun, Une procédure d'apprentissage pour réseau a seuil asymmetrique. Proc. Cogn. **85**, 599–604 (1985)
19. V. Lyovin, *An Introduction to the Languages of the World* (Oxford University Press, New York, 1997)

20. F. Magdoff, J.B. Foster, *What Every Environmentalist Needs to Know About Capitalism: A Citizen's Guide to Capitalism and the Environment* (Monthly Review Press, New York, 2011)
21. W. McCulloch, W. Pitts, A logical calculus of ideas immanent in nervous activity. Bull. Math. Biophys. **5**(4), 115–133 (1943)
22. M. Minsky, S. Papert, *An Introduction to Computational Geometry* (MIT Press, Cambridge, 1969)
23. D.B. Parker, Learning-logic. Technical report-47, MIT Center for Computational Economics, Cambridge (1985)
24. F. Rosenblatt, The perceptron: a probabilistic model for information storage and organization in the brain. Psychol. Rev. **65**(6), 386–408 (1958)
25. I. Rubenstein, Regulating privacy by design. Berkeley Technol. Law J. **26**, 1409 (2012)
26. S. Russell, P. Norvig, *Artificial Intelligence: A Modern Approach* (Pearsons, Harlow, 2009)
27. R. Schiller, *Irrational Exhuberance* (Princeton University Press, New York, 2000)
28. V.M. Schonberger, K. Cukier, *Big Data: A Revolution That Will Transform How We Live, Work and Think* (Houghton Mifflin Harcourt Publishing Company, New York, 2013)
29. P. Singer, *Animal Liberation: A New Ethics for Our Treatment of Animals* (New York Review of Books, New York, 1975)
30. M.W. Vail, J.B. Earp, A.I. Anton, An empirical study of consumer perceptions and comprehension of web-site privacy policies. IEEE Trans. Eng. Manage. **55**(3), 442–454 (2008)
31. P.J. Werbos, *Beyond Regression: New Tools for Prediction and Analysis in the Behavioral Sciences* (Harvard University, Cambridge, 1975)
32. T. Wilson, J. Wiebe, P. Hoffman, Recognizing contextual polarity in phrase-level sentiment analysis, in *Proceedings of the Conference on Empirical Methods in Natural Language Processing (EMNLP-2005)* (2005)
33. J.D. Young, Commitment analysis to operationalize software requirements from privacy notices. Requirement Eng. **16**(1), 33–46 (2010)
34. X. Zhang, Y. LeCun, *Text Understanding from Scratch* (2016). arXiv:1502.01710

Intelligent Decision Support System for an Integrated Pest Management in Apple Orchard

T. Padma, Shabir Ahmad Mir and S.P. Shantharajah

Abstract Sizable human population of the world is associated directly or indirectly with the agriculture and natural resources found on the earth and is important to the study of sustainability. Despite of difficulties in practical agriculture, new opportunities have been created by sophisticated technology-driven change to address judicious use and management related problems of resources and thereby improving human wellbeing. In agriculture, Intelligent Decision Support Systems (IDSS) have been used for optimization of number of planning and decision making challenges under variable number constraints based on noisy data. This research describes an IDSS to implement and optimize pest and disease protection decision making processes within temperate regions of India; develops hybrid algorithm using Case Based Reasoning and Database Technology and implement the same using web based client server architecture. The accuracy of decision making process provided by the system has been 90.20% and it can provide significant support to the apple farmers in decision making towards Eco-friendly pest management practices.

Keywords Integrated pest management · Case based reasoning · Insect pest · Data analysis · Behavioral patterns · Computational intelligence

T. Padma (✉)
Sona College of Technology, Salem, Tamilnadu, India
e-mail: padmatheagarajan@gmail.com

S.A. Mir
Sher-e-Kashmir University of Agricultural Sciences and Technology of Kashmir,
Jammu, India
e-mail: mirshabir972@gmail.com

S.P. Shantharajah
VIT University, Vellore, Tamilnadu, India
e-mail: shantharajah.sp@vit.ac.in

© Springer International Publishing AG 2017   225
A.K. Sangaiah et al. (eds.), *Intelligent Decision Support Systems for Sustainable Computing*, Studies in Computational Intelligence 705,
DOI 10.1007/978-3-319-53153-3_12

1 Introduction

In India, apple fruit production has become main horticultural activity in the northern states of Jammu and Kashmir, Himachal Pradesh and Uttarkhand, where this industry is considered as a backbone of economy. An area of 289 thousand hectares are under this fruit with an annual production of about 2890.6 thousand tones [1]. These regions have more than 30 million human population and enormous natural resources, and hence are important to the study of sustainability. Moreover, the demand for fresh fruits especially apple has shown faster growth than other field crops in the recent times [2]. India figures among top ten apple producing countries and its contribution to the overall production is 3% [3–5]. However, the production and quality of apple is poor as compared to that of the developed countries because of several factors including insect and diseases, which cause great loss to the apple growers annually [6–9]. In order to minimize this loss, stakeholders of horticulture sector in developing countries like India are implementing different integrated pest and disease prevention programs for various fruits crops including apple. Implementation of these programmes requires complex tactical decisions and integration of information from multiple sources. Integrated Pest Management (IPM) programmes are often formulated without identifying the real needs of farmers and their advisors wherein imprecise and interconnected information is communicated in the form of bulletins and/or advisories. Due to ever-changing weather conditions and integration of information from multiple sources together with complexities associated with the decision process of IPM, the extension systems are successful only when supported with intelligent systems to address complex decision making. This imprecise and uncertain nature of the IPM includes information that is vague, fragmentary, not fully reliable, contradictory, deficient, and overloading; thus affecting quality fruit production of apple besides it is costly, and can be detrimental to the environmental as well. Intelligent system based decision support can improve decision making process by providing timely solution to critical problems, increase productivity and decrease costs under complex environment [10, 11]. It is surprising that computational methods have not been applied for providing better planning and decision making solutions in these regions. Therefore, due to lack of fast, accurate, reliable and weather aware decision support with the extension system, extension bulletins and advisories often becomes ineffective. This imprecise nature of the IPM is effecting quality fruit production of apple in the temperate regions suitable for apple production. Mathematical modeling is inadequate in these cases given imprecise nature of the IPM problem due to the complex, uncertain and stochastic nature of processes to be modeled. Despite of these difficulties, technology driven change in the form of sustainable computational intelligence can create new opportunities to address poor management of apple production resources thereby improve quality of life.

Computational Intelligence (CI) techniques use computational power of computers to integrate, analyze and share large volume of noisy data in real time, using diverse analytical techniques to discover important information suitable for better decision making. CI techniques can move forward farm economics leaps and bounds by transformation of theoretical concept of computational intelligence techniques into practical agriculture. The process has already begun by the development of weather forecasting [12], crop health monitoring [13], detection and classification of plant leaf diseases [14], fruit and grain grading system [15, 16], precision farming [17], crop prediction [18] and estimation of morphological parameters and quality of agricultural products [19] and many others are clear examples of how these techniques can be future endeavor of farming industry. Application of these techniques can reduce severity of disease and pathogens problems in apple orchards while minimizing applications of spraying substances by focusing on life cycles of pathogens and weather conditions, thus boosting the ability of agricultural and environmental conditions to continue support life on earth.

Considering the limited nature of information presented in the pest protection programme; vagaries in environmental and climatic conditions as well as growing resistant behavior of pathogens towards spraying substances, IPM programme could not achieve desired results. Therefore, in order to supplement climatic and pest phonological data to IPM module, computational intelligent DSS have been developed for temperate regions of India to predict likely consequences of presence of insects, pests or disease in the apple orchards. The effort has resulted in adoption of efficient management practices to minimize repetitive sprays, environmental impact and risk of workers as well as cost of spraying.

This research describes an intelligent DSS that was developed to help apple growers within temperate regions of India in decision making about IPM based on the principles of sustainability to enhance the economic viability of apple fruit crop through optimized management of insect, pest and disease protection with reduction in spraying substance usage. A Hybrid computational intelligent algorithm based on Case Based Reasoning (HCBR) and database technology is presented which supports retrieval, reuse and revise IPM knowledge to provide optimal decision support for various spraying substance viz-a-viz disease diagnosis. The algorithm is implemented using web based client-server system, which provide decision making and predictive forecast support related to pathogens and diseases of apple.

2 Background

Food production and demand in some of the countries like India are fragile due to changes in climatic conditions, spread of diseases and constant degradation of farmland. It is very useful to anticipate the threats to sustainability by predicting climatic conditions; spread of disease or pathogens and shrinking farmland size. But the assessment process of these factors is not straight forward using conventional methods. Besides, combined approach of these factors makes this process expensive, time consuming and inadequate due to scarcity of suitable extension staff, difficulty of logistic transport and timely paper report coordination.

On the other hand, IPM is a synergy of diseases and environmental conditions. Diseases cause severe damage to the tree and eventually economic loss to the farmers. Disease diagnosis in earlier stage is desirable. But it needs study of numerous characteristics of plant, environment and behavior of pathogens and diseases, which are not distinguishable in most cases. Besides, identification and presence of disease is prerequisite for its treatment. Therefore, identification and treatment using computational intelligent techniques is a must and it finds great significance in precision disease management.

3 Integrated Pest Management (IPM)

Integrated Pest Management (IPM) is an effective and environmentally sensitive approach towards pest management that uses a combination of commonsense practices. IPM programs use current, comprehensive information on the life cycles of pests and their interaction with the environment. This information is combined with available pest control methods to manage pest damage using most economical means together with the least possible risk to people, property, and the environment.

All insects and other living organisms do not require management. Many of them are injurious while several others are also beneficial. IPM programs monitors and identifies pest infestation precisely, so that appropriate control decisions can be made in colligation with action thresholds. This monitoring and identification process removes the possibility of imprecise application of pesticides and their timing.

Conjugative approach of field monitoring, disease or pest identification, and action thresholds resolves requirement of pest control. Proper management methods are accordingly evaluated and suggested under IPM programs taking into consideration the effectiveness and risk of the method to be used. At first, effective and less risky pest controls like pheromones to disrupt pest mating or mechanical control such as trapping. This may followed by highly targeted control like usage of spraying substances, when monitoring, identifications and action thresholds indicate less risky method as non-effective. Broadcast spraying of non-specific pesticides is the last resort.

4 Computational Intelligence

Due to design of natural and biological algorithmic models for complex problems, artificial intelligent (AI) systems are being developed. CI is a sub-branch of AI or Intelligent systems, which is a fairly new research field with competing definitions. According to [20] it is the branch of science studying problems for which there are no effective computational algorithms. Computational Intelligence techniques mainly focus on *strategy* and *outcome* by putting emphasis on heuristic algorithms such as fuzzy systems, evolutionary computation and neural networks. They are used to address complex real-world problems to which application of mathematical modeling is inadequate due to the complex, uncertain and stochastic nature of processes to be modeled [21]. In essence, CI refers to the ability of a computer to learn a specific complex task from data or observations. CI methods can mimic human's way of reasoning to reasonable extent, using non exact and non-complete knowledge, to produce control actions in an adaptive way.

CI is based on fuzzy logic, case-based reasoning, neural networks, evolutionary computation, learning systems, probabilistic methods, nature-inspired systems, artificial immune systems, artificial neural networks and swarm intelligence. CI techniques facilitate intelligence in ever changing as well as composite environments through adoptive learning of mechanisms possessed by the system under study. For adoptive learning of the mechanism, we often use Artificial Intelligence (AI) paradigms that demonstrate an ability to gain insight or adapts to new situations through discovery, abstraction, association and generalizations. These techniques have been successfully applied to solve real-world problems individually. But the present trend is to develop hybrid CI paradigms, since no one paradigm can be used in all situations.

Sustainability is an ability to produce such a large quantity of energy which is enough to support biological and natural systems on earth. In the context of Computational Intelligence (CI), it is the computational power of computers to process complex data using sophisticated mathematical models, analytical techniques and fuzzy inferences. In agriculture, it attempts to optimize economic, environment and social resources to boost ability to produce enough food for human and animal consumption. Use of sustainable computing in agriculture can aid towards this endeavor using mathematical models and computer technology, focusing action taken and decision making for better economic gains viz-a-viz environmental protection.

4.1 Case Based Reasoning

Case-based reasoning is a four step approach to problem solving that draws attention to the role of prior experience during future problem solving. In other words, new problems are solved by reusing of prior proven intelligence and if necessary adapting the solutions to similar problems that were solved in the past [22]. This technique is a relatively newcomer to CI, which arose out of research into cognitive science, most conspicuously that of Schank [23–25]. Its basis was stimulated by a desire to understand how people remember information and are in turn reminded of information. Subsequently, it was recognized that people commonly solve problems by remembering how they solved similar problems in the past [26]. Conceptually CBR is usually represented by CBR-cycle as shown in Fig. 1 given by Aamodt and Plaza [27].

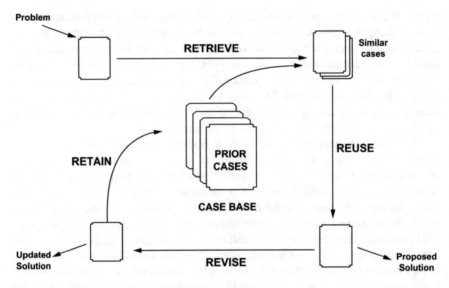

Fig. 1 The CBR cycle

This cycle comprises four activities popularly known as the 4RE's:

Retrieve similar cases to the problem description. This retrieve phase regains relevant cases form memory pertinent to the given problem. A case consists of a problem, its solution, and procedure about how the solution was derived. For example, IPM for apple fruit; here a farmer can recall most relevant experience using which he successfully controlled pathogens, together with the procedure he followed with justifications for decisions made along the way, constitutes IPM retrieved case.

Reuse a solution suggested by a similar case. Reuse phase maps the solution from the previous case. To the IPM problem, which may involve adapting the solution as needed to fit the new situation.

Revise that solution to better fit the new problem. After mapping the previous solution to the target situation, the revise stage tests the new solution, here in the context of IPM problem in hand. It may revise the procedure as per the current situation. In the perspective of IPM for apple, suppose there is a retained case for the control of sooty blotch with the recommendation, "Application of Mencozeb 75WP (75 g) per 100 L of water". The recommendation will be retrieved from the memory along with weather forecast, which suggests likely rainfall in next 24 h. Therefore, the case will be revised with the addition of "Spray with stickers or spray day after tomorrow".

Retain the new solution once it has been confirmed or validated. After successful adoption of the solution to the target problem, the resulting experience is stored in memory as a new case.

The guiding principle behind CBR is solving a problem explicitly through reusing a solution from a similar past problem. It retrieves cases from a case-library based on the principle of similarity of cases in the library to the current problem description. Then it attempts to reuse the solution suggested by a retrieved case with or without any revision. Eventually CBR systems increase its knowledge by retaining new cases.

4.2 CBR Techniques

CBR can be implemented using Nearest neighbour technique, Induction technique, Fuzzy logic technique and Database technology [27].

Nearest neighbor. Nearest neighbour finds case to case similarity of the problem in the case-library and determine attribute for each case. Then the measure is multiplied by a weighting factor. A measure of the similarity of that case in the library to the target case is provided by calculating sum of the similarity of all attributes. The process is represented by the Eq. (1).

$$Similarity(T, S) = \sum_{k=1}^{n} f(T_k, S_k) \times w_k \qquad (1)$$

where T is the target case; S the source case; k the number of attributes in each case; f a similarity function for attribute k in cases T and S; and w the importance weighting of attribute k.

Similarities are usually normalized to fall within a range of 0–1 (where 0 is totally dissimilar and 1 is an exact match). The normalization can also be indicated as a percentage, where 0% would mean dissimilar and 100% an exact match. The use of nearest neighbour technique is well illustrated by the Wayland system [28].

Induction. Inductive retrieval algorithm is a technique that determines which features do the best job in discriminating cases and generates a decision tree type structure to organize the cases in memory [29]. This approach is very useful when a single case feature is required as a solution, and when that case feature is dependent upon others.

These techniques are commonly used in CBR. Induction algorithms, such as ID3 generates decision trees from case histories. All induction algorithms identify patterns amongst cases and partition the cases into clusters. Each cluster contains cases that are similar with the case in hand. Definition of target case is required in induction. Induction algorithms are importantly being used as classifiers to cluster similar cases together, wherein it is assumed that cases with similar problem descriptions will have similar solutions.

Fuzzy logic. Fuzzy logic is a way of formalizing the symbolic processing of fuzzy linguistic terms, such as excellent, good, fair and poor, which are associated with differences in an attribute describing a feature [22]. In fuzzy logic, number of linguistic terms is not restricted. Fuzzy logic intrinsically represents notions of similarity, since the linguistic values 'good' and 'excellent' are closer to each other than 'poor'. In order to find a solution to a problem, fuzzy preference function based on CBR can be used to calculate the similarity of a single attribute of a case with the corresponding attribute of the target.

Database approach. At its simplest form, CBR could be implemented using database technology. Databases are efficient means of storing and retrieving large volumes of data. If problem descriptions could make well-formed queries it would be straight forward to retrieve cases with matching descriptions. A problem along with using database technology for CBR is that retrieval of database using exact matches to the queries.

This is commonly augmented by using Wildcards, such as "WESTp" matching on "WESTMINSTER" and "WESTON" or by specifying ranges such as " , 1965". The use of Wildcards, Boolean terms and other operators within queries may make a query more general, and thus more likely to retrieve a suitable case, but it is not a measure of similarity. However, by augmenting a database with explicit knowledge of the relationship between concepts in a problem domain, it is possible to use SQL queries and measure similarity [27].

Hybrid approach. Hybrid approach uses more than one technique as described above to solve diverse problems. Similarly, the earlier studies [30–32] have addressed the hybrid fuzzy approaches for intelligent decision making process.

5 Towards Computational Intelligent DSS for IPM

Application of CBR to IPM is quite relevant considering different representation cases, requirement of past experience and wide range of possible responses. This section presents an intelligent computational solution based on the principles of CBR methodology using database approach which will be developed and implemented on World Wide Web. CBR has been selected due to its ability to consider past experience, provides representative cases which are similar to current problems, and provides solutions which take into consideration a range of possible responses.

An intelligent component to support decision-making is able to integrate existing phonological data related to pests and current weather conditions to support pest population predictions at economic and injury threshold levels. Beyond economic threshold level appropriate action as per the package of practices issued by the competent authorities shall be performed automatically by the system. As IPM requires past experience, efforts have been made where the system can learn from new situations and contributing to timely delivering interventions that benefit intended growers.

5.1   *Computational Intelligence Techniques Used*

In this research, induction technique and database approach are used together by extracting their relative potentials to address IPM of apple. The approach is used keeping in view unstructured nature of the knowledge base of IPM. Partitioning feature of induction technique and matching capability of database approach has been combined to get a CBR algorithm best suited for pest management problem. This study develops IPM intelligent agent comprising of two modules where the main interface is as shown in Fig. 2.

First module addresses the periodicity of different sprays as per the phonological stages of apple tree keeping in view current weather conditions. For which a similar case like the one in hand is retrieved from the database using MYSQL's data manipulation command *Select*. The case is then partitioning into clusters using *wildcards*.

Firstly, using *select* most appropriate or similar case is identified from the multi-lingual knowledge base. Due to bi-lingual and unstructured data, wildcards are used to partition the data into desirable segments so that the recommendations will be organized into a most interpretive manner. The algorithm used for this purpose is given hereunder:

Retrieve *description* from knowledge-base where *case* = present phonological stage;

Fig. 2 Main interface of intelligent agent for apple IPM

Partition *description* using *wildcard* (^) based on languages to get *Partition [0]* and *Partition [1]*;

Based on language option **Partition** *Partition [0]* using
wildcard (1) to get *Recommendations* and *Warnings* in the said
language;

Partition *Recommendations* to get *segmented* options of
Spray;

And **Partition** *Warnings* to get *segmented* options of
Warning;

Reuse *Recommendations*;

Reuse *Warnings*;

Revise *description* with the present and future predicted
weather conditions;

Retain *description* for future use;

Second module performs symptom based disease diagnosis, wherein based on
one or more symptoms, similar case is retrieved from the multi-lingual knowledge
base using *Select* command of MYSQL's data manipulation. For multiple symptoms, the revision is done and the same is retained for future use. Here, new
symptoms if any are added to the image database.

The case is portioned in the form of disease introduction, life cycle and control
measures based on the language option selected by the user. The CBR algorithm
used for this purpose is as under:

Retrieve *description* from knowledge-base where *case* is
similar to symptom(s);

Partition *description* using *wildcard* (^) based on languages to get *Partition [0]* and *Partition [1]*;

Based on language option **Partition** *Partition [0]* using
Wildcard (1) to get *introduction, life cycle and control
measures* in the said language;

Reuse *descriptions*;

Revise *control measures* with the present and future predicted weather conditions;

Retain *description* for future use;

5.2 Description of the Intelligent IPM DSS

The overall working of the system is divided into two modules viz., Spray schedule
and Disease diagnosis.

Spray schedule module. This module gives recommendations for chemical
sprays based on the phonological stage of apple tree and weather condition.
Fourteen phonological stages of apple fruit development such as Dormant, Silver

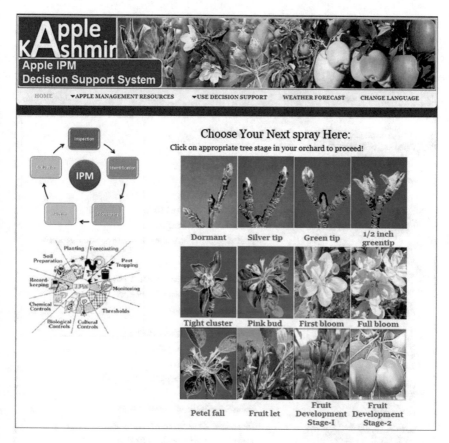

Fig. 3 Phenology based spray schedule

tip, Green tip, Half inch green tip, Tight cluster, Pink bud, First bloom, Full bloom, Petal fall, Fruit let and four stages of actual fruit development. This spray schedule interface is as shown is Fig. 3.

While user clicks on the appropriate stage, the system applies the algorithm stated already to get the next spray to be used in the orchard. For illustration, assuming that the user chooses Petal fall stage, the system generates spray recommendations for that stage with the warnings as shown in Fig. 4. Here stages are assumed having different cases. Users can reuse information unless the knowledge base will be modified and retained with the findings of new research.

Disease diagnosis module. This module helps to diagnose diseases pertaining to apple pests and diseases based on symptoms, which have been incorporated in the form of pictures categorized under different parts of the tree. Here user first selects the part in which disease has been noticed to emerge using the interface shown in Fig. 5. This has done to minimize the search space to improve response time. Then disease symptoms pertaining to that part are displayed, wherein user can select one

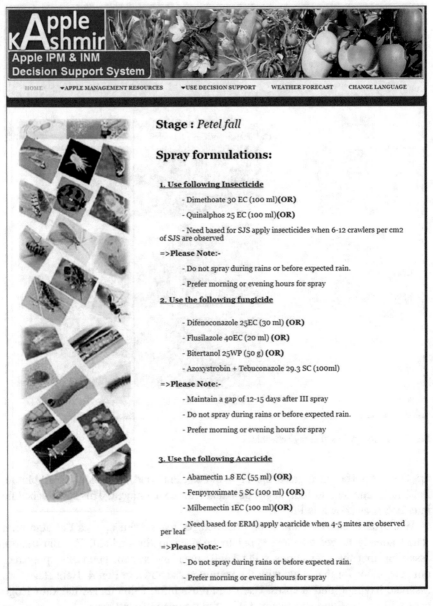

Fig. 4 Recommendations for petal fall stage

or many symptoms based on their observations in the field. The system retrieves similar case from the knowledge base, based on the algorithm given already. The retrieved information can be reused based on the symptoms. Knowledge base can be modified by the domain expert whenever new recommendations for a particular

Fig. 5 User Interface for symptom based disease diagnosis

disease emerge. Finally the case is retained in the knowledge base for future use. For illustration, assuming user notices some symptoms on the fruits and chooses fruit part as diseased part using the interface shown in Fig. 5. Then the system generates suitable case for that part in the form of indexed pictures indicating symptoms as shown in Fig. 6. Here let's assume that the user choose picture indexed by number thirteen (13) as the symptom, then the system retrieves case with similar case description and generates recommendations automatically keeping in view present phonological stage with the warnings if any. Users can reuse this information unless the knowledge base will be modified and retained with the findings of new research. Assuming that the present phonological stage is peanut stage, the recommendations generated by the system is shown in the Fig. 7.

6 Results and Discussion

Design, development and evaluation of agricultural DSSs have widely been documented in literature [33–35]. However design and development of intelligent IPM DSS follows as related research [36–39], in which knowledge base design

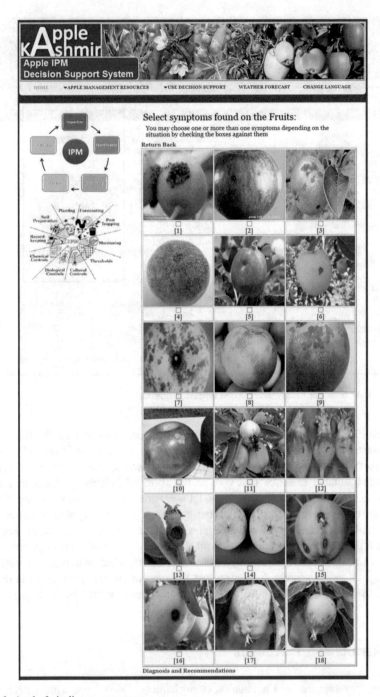

Fig. 6 Apple fruit disease symptoms

Fig. 7 Recommendations generated for fruit worm at peanut stage

comprises 15 pathogens already identified by the domain experts pertaining to apple prevalent in the temperate regions of India (Table 1). Technically, for better uptake of any agricultural DSS, it is delivered on the web; provides an easy and complete user-friendly interface; enables continuous and flexible access; automate and integrate data from multiple sources; helps in decision-making and not attempts to replace decision-maker; use validated models and involve all stakeholders [35]. Therefore, client-server web-architecture with hypertext manipulation language and the hypertext preprocessor PHP as front end and MySQL as backend were used so that the system exhibits continuous and flexible build–in dynamic query access; automate and integrate weather data from multiple nearby sources and can be accessed by all regional apple fruit growers. Besides, easy interface with pictorial inputs were designed for the end users. Every effort was made to use validated models and involve all stakeholders in the development process.

For usefulness and adoption, outcome is very important aspect of any DSS [39]. Research studies also envisage that users should be satisfied with the outcome of DSS [40]. In order to study the outcome of DSS with respect to sampling

Table 1 Details of
pathogens modeled in the
system

Common name	Scientific name
San Jose scale	*Quadraspidiotus perniciosus*
Woolly apple aphid	*Eriosoma lanigerum*
European red mite	*Panonychus ulmi*
Codling moth	*Cydiapomonella*
Powdery Mildew	*Podosphaerea leucotricha*
Root rot	*Dematophora necatrix*
Cankers	*Botryosphaeria spp.*
Apple Mosaic	*Apple Mosaic Virus*
Leaf and fruit spot	*Entomosporium maculatum*
Marssonina blotch	*Marssonina coronaria*
Sooty blotch	*Gloeodes pomigena*
Fly speck	*Schizothyrium pomi*
Apple Scab	*Venturia inaequalis*
Alternaria leaf blotch	*Alternaria mali*
Collar rot	*Phytophthora cactorum*

procedure, 383 trial cases involving domain experts comprising of Entomologists and Plant were conducted. The maximum scores that an evaluator could obtain was 100% (If all decisions made by the systems satisfies him/her) and the minimum score that a respondent could obtain depending upon number of satisfied decisions minus number of un-satisfied decisions. The summary of trial cases and accuracy of the decisions are indicated in the Table 2.

The trial cases conducted for judging efficiency of the system were of two types of decisions supported by the system viz., spray recommendations and disease diagnosis. For spray related decisions, 10 trail-cases were evaluated by all experts. The ten trial cases were different identified phonological stages for which sprays and their compositions were recommended. Concept of retrieve, reuse, revise and retain was demonstrated so that processed information can be retrieved and reused until it is applicable. Once new findings in the form of sprays and their formulations are found, knowledge base could be revised and retained accordingly. For spray decision process, fruit development stages were divided into 12 phonological stages from dormancy to post harvest stages. These stages were: dormancy, silver tip, green tip, pink bud, peanut, fruit development stage-1, fruit development stage-2, fruit development stage-3 and fruit development stage-4. Under these categories different recommended sprays for pathogens were collected and their associated knowledge base was developed. The only input demanded by the system is selection of the stage. The system generates probable stage(s) which are subject to validated by the users. The results of the trial cases suggest that 100% accuracy was found in the recommendations made by the system pertaining to the 10 user cases defined by spray recommendations.

However, for evaluation of disease diagnosis module, varied number of trail cases was performed by the experts due to diverse expertise of the experts as well as

Table 2 Summary of trial cases and accuracy of the decisions made by the system

Experts	Spray recommendations (N = 100)				Disease diagnosis (N = 283)				Overall
	Use-case	No. of accurate decisions	No. of inaccurate decisions	Accuracy (in %)	Use-case	No. of accurate decisions	No. of inaccurate decisions	Accuracy (in %)	accuracy (in %)
E-1	10	10	–	100	26	24	2	92.31	96.16
E-2	10	10	–	100	32	26	6	81.25	90.63
E-3	10	10	–	100	35	34	1	97.14	98.57
E-4	10	10	–	100	29	28	1	96.55	98.28
E-5	10	10	–	100	41	38	3	92.68	96.34
E-6	10	10	–	100	22	22	0	100.00	100.00
E-7	10	10	–	100	17	15	2	88.24	94.12
E-8	10	10	–	100	38	35	3	92.11	96.06
E-9	10	10	–	100	25	20	5	80.00	90.00
E-10	10	10	–	100	18	16	2	88.89	94.45
Total	100	100	–	100	283	258	25	90.92	95.46

multitude of the diseases and pests together with diversified nature of symptoms for a disease and disorder found in the apple. The trial cases were performed based on the expertise and availability of time with the experts. For disease diagnosis process, apple disease, disorders and pathogens were divided into six categories based on the different parts of the trees. These categories were: root, stem, branches, leaves, flowers and fruits. Under these categories varied number of diseases, pests and disorders known to exist in the study area were identified and their associated knowledge base developed.

The knowledgebase comprised of introduction, symptoms, life cycle and remedial measures were implemented so that farmers as well as students can get benefitted. The results of the evaluations suggest that minimum 17 user cases were evaluated by the Expert 'E-7' and maximum user case of 41 were evaluated by the Expert 'E-5'. Maximum accuracy (100%) was reported by Expert 'E-6' with 22 user cases. While as minimum accuracy of 80% was reported by the Expert 'E-9'. Overall, for disease diagnosis, 283 user cases were performed in which 258 cases were found accurate and 25 decisions were found inaccurate. The overall accuracy was 95.46%. The nature of inaccuracy exhibited by the intelligent decision support system has been documented and it shall pave the way forward to improve the performance of the system.

7 Scope, Significance and Limitations of the Study

The computational algorithms applied in this research can be used for the development of computationally intelligent IPM for the other crops. In this study, the algorithm was used to provide optimum spray decisions, disease identification and treatment pertaining to the apple, which finds great significance in precision disease management research. However, only a limited number of pathogens have been modeled. Besides the algorithm presented hereby can work efficiently on a suitable database schema, wherein different wildcards will mean different things for the algorithm.

8 Conclusions

In this chapter, hybrid case based reasoning computational intelligence techniques have been presented for pest management in apple. The technique uses four stage process of case based reasoning methodology viz., retrieve, reuse, revise and retain. The retrieve process is implemented in association with the matching logic performed on the knowledge base. The knowledge is presented in suitable format for the end user who reuses the same until the same is revised by the domain experts to incorporate latest research findings. The retain process is incorporated by storing the knowledge in the knowledge base. For sustained development of apple production, computationally intelligent techniques based decision support system has a role to play, in order to handle knowledgeably the vague multifaceted nature of IPM. The proposed Intelligent IPM decision support agent reduces the complexity of decision making to workable dimensions.

References

1. GoI Area, Production and Productivity of Various Fruits in India. Ministry of Agriculture and Cooperation, Government of India 2013, New Delhi
2. N.A. Rather, Parvaze Ahmad Lone, A.A. Reshi, M.M. Mir et al., An analytical study on production and export of fresh and dry fruits in Jammu and Kashmir. Int. J. Sci. Res. Publ. **3** (2), 1–7 (2013)
3. AESA based IPM package for Apple. Department of Agriculture and Cooperation, Ministry of Agriculture, Government of India 2014, p. 50
4. Jerneja Jakopic, Anka Zupan, Klemen Eler, Valentina Schmitzer, Franci Stampar, Robert Veberic et al., It's great to be the King: apple fruit development affected by the position in the cluster. Sci. Hortic. **194**, 18–25 (2015)
5. M.L. Corollaro, I. Endrizzi, A. Bertolini, E. Aprea, M.L. Demattè, F. Costa, F. Biasioli, F. Gasperi et al., Sensory profiling of apple: methodological aspects, cultivar characterisation and postharvest changes. Postharvest Biol. Technol. **77**, 111–120 (2013)
6. A. Sherwani, M. Mukhtar, A.A. Wani, *Insect Pests of Apple and their Management, Insect Pests Management of FRUIT CROPS* (Biotech Books, 2016)
7. L. London, S. De Grosbois, C. Wesseling, S. Kisting, H.A. Rother, D. Mergler et al., Pesticide usage and health consequences for women in developing countries: out of sight out of mind? Int. J. Occup. Environ. Health **8**(1), 46–59 (2002)
8. J. Bonany, A. Buehler, J. Carbó, S. Codarin, F. Donati, G. Echeverria, S. Egger, W. Guerra, C. Hilaire, I. Höller, I. Iglesias, K. Jesionkowska, D. Konopacka, D. Kruczyńska, A. Martinelli, C. Pitiot, S. Sansavini, R. Stehr, F. Schoorl et al., Consumer eating quality acceptance of new apple varieties in different European countries. Food Qual. Prefer. **30**(2), 250–259 (2013)
9. M.S. Ahmad, A. Nayyer, A. Aftab, B. Nayak, W. Siddiqui et al., Quality prerequisites of fruits for storage and marketing. J. Postharvest Technol. **2**(1), 107–123 (2014)
10. T. Wascinski, A. Michalczyk et al., Agricultural business extension aided by the case-based reasoning method. Ann. Warsaw Univ. Life Sci. SGGW Agric. **52**, 107–114 (2008)
11. D. Ruan (eds.), *Atlantis Computational Intelligence Systems*, in Computational Intelligence in Complex Decision Making Systems (World Scientific, Atlantis Press, 2010)
12. N.O. Bushara, A. Abraham et al., Computational intelligence in weather forecasting: a review. J. Netw. Innov. Comput. **1**, 320–331 (2013)
13. S. Arivazhagan, R.N. Shebiah, S. Ananthi, S.V. Varthini et al., Detection of unhealthy region of plant leaves and classification of plant leaf diseases using texture features. Agric. Eng. Int. **15**(1), 211–217 (2013)
14. A.M. Vyas, B. Talati, S. Naik et al., Colour feature extraction techniques of fruits: a survey. Int. J. Comput. Appl. **83**(15), 15–22 (2013)
15. S.P. Shouche, R. Rastogi, S.G. Bhagwat, K.S. Jayashree et al., Shape analysis of grains of Indian wheat. Comput. Electron. Agric. **33**, 55–76 (2001)

16. C. Bauckhage, K. Kersting, *Data Mining and Pattern Recognition in Agriculture* (Springer, Berlin, 2013)
17. J.R. Prasad, P.R. Prakash, S.S. Kumar et al., Identification of agricultural production areas in Andhra Pradesh. Int. J. Eng. Innov. Technol. **2**(2), 137–140 (2012)
18. A. Khoshroo, A. Keyhani, S. Rafiee, R. Zoroofi, Z. Zamani et al., Pomegranate quality evaluation using machine vision, in *Proceedings of the First International Symposium on Pomegranate and Minor Mediterranean Fruits*, 2009, pp. 347–352
19. D. Wlodzislaw, M. Jacek (eds.), *Challenges for Computational Intelligence* (Springer, 2007), p. 63
20. C.K. Riesbeck, R. Schank, *Inside Case-Based Reasoning* (Erlbaum, Northvale, NJ, 1989)
21. S. Adeli, N. Hojjat, *Computational Intelligence: Synergies of Fuzzy Logic, Neural Networks and Evolutionary Computing* (Wiley, New York, 2013)
22. B. López, *Case-Based Reasoning: A Concise Introduction, Synthesis Lectures on Artificial Intelligence and Machine Learning* (University of Girona, Spain, 2013)
23. K.J. Hammond, *Case-Based Planning: Viewing Planning as a Memory Task* (Academic Press, New York, 2012)
24. R.C. Schank, R.P. Abelson, Scripts, plans, goals, and understanding: an inquiry into human knowledge structures. Artif. Intell. Ser. (2013)
25. R. Schank (ed.), *Dynamic Memory: A Theory of Learning in Computers and People* (Cambridge University Press, New York, 1982)
26. I. Watson, Case-based reasoning is a methodology not a technology. Knowl. Based Syst. **12** (5–6), 303–308 (1999)
27. A. Aamodt, E. Plaza et al., Case-based reasoning: foundational issues methodological variations and system approaches. AI Commun. **7**, 39–59 (1994)
28. C.J. Price, I.S. Pegler (eds.), *Deciding Parameter Values with Case-Based Reasoning*. Progress in Case-based Reasoning. Lecture notes in artificial intelligence, vol. 1020 (Springer, 1995), pp. 121–133
29. I. Watson, *Applying Case-based Reasoning: Techniques for Enterprise Systems* (Morgan Kaufmann, CA, USA, 1997)
30. O.W. Samuel, G.M. Asogbon, A.K. Sangaiah, P. Fang, G. Li, An integrated decision support system based on ANN and Fuzzy_AHP for heart failure risk prediction. Expert Syst. Appl. **68**, 163–172 (2017). Elsevier Publishers
31. A.K. Sangaiah, A.K. Thangavelu, X.Z. Gao, N. Anbazhagan, M.S. Durai, An ANFIS approach for evaluation of team-level service climate in GSD projects using Taguchi-genetic learning algorithm. Appl. Soft Comput. **30**, 628–635 (2015)
32. A.K. Sangaiah, A. K. Thangavelu, An adaptive neuro-fuzzy approach to evaluation of team-level service climate in GSD projects. Neural Comput. Appl. **23**(8) (2013) (Springer Publishers). doi:10.1007/s00521-013-1521-9
33. P.S. Carberry, Z. Hochman, R.L. McCown, N.P. Dalgliesh, M.A. Foale, P.L. Poulton, J.N.G. Hargreaves, D.M.G. Hargreaves, S. Cawthray, N. Hillcoat, M.J. Robertson et al., The FARMSCAPE approach to decision support: farmers', advisers', researchers' monitoring, simulation, communication and performance evaluation. Agric. Syst. **74**, 141–177 (2000)
34. F.P. de Vries, P. Teng, K. Metselaar (eds.), Systems approaches for agricultural development, in *Proceedings of the International Symposium on Systems Approaches for Agricultural Development*, Bangkok, Thailand, 1993
35. V. Rossi, P. Meriggi, T. Caffi, S. Giosué, T. Bettati (eds.), *A Web-based Decision Support System for Managing Durum Wheat Crops. Advances in Decision Support Systems* (INTECH, Croatia, 2010)
36. J.D.C. Little, Models and managers: the concept of a decision calculus. Manage. Sci. **16**(8), 1970 (1970)
37. J.H. Moore, M.G. Chang et al., Design of decision support systems. Data base **12**, 1–2 (1980)
38. Z. Tagir, I. Ibragimov, T. Ibragimova, S. Sergey et al., *Decision Support Systems for Cereal Crop Disease Control* (EFITA/WCCA, Vila Real, Portugal, 2005)

39. S.A. Mir, I.A. Mir, S.M.K. Quadri, N. Ahmad, S. Angchuk et al., ONVAREF: a decision support system for onion varietals reference. Afr. J. Agric. Res. **8**(48), 6275–6282 (2013)
40. A.M. Gil-Lafuente, C. Zopounidis (eds.), Decision making and knowledge decision support systems, in *VIII International Conference of RACEF,* Barcelona, Spain, 2015

Analysis of Error Propagation in Safety Critical Software Systems: An Approach Based on UGF

R. Selvarani and R. Bharathi

Abstract Designing fault free software systems becomes an essential practice towards Safety Critical Software System (SCSS) manufacturing. The error free scenario of SCSS will support the systems perfect functioning. The proposed approach is based on universal generating function to compute the error inclusion in the output of the selected safety critical system. This paper presents an Error Propagation Metric called Safety Metric SM_{EP}, which can be characterized depending on the performance rate of the software module. Through this, the performance distribution of system modules and the system with respect to safety metric SM_{EP} is quantified.

Keywords Error · Safety metric · Performance · Modules · Subsystem · System

1 Introduction

SCSS may contain faults, although the system has been well tested, used and documented. If one part of a system fails, this can affect other parts and in worst-case results in partial or even total system failure. To avoid such incidents research on failure analysis is of high importance. A failure of a safety critical system can be defined as "the non performance or incapability of the system or a component of the system to perform or meet the expectation for a specific time under stated environmental conditions". Error propagation between software modules is a qualitative factor that reflects on the reliability of a safety critical software product.

R. Selvarani
Computer Science and Engineering, Alliance University, Bangalore, India
e-mail: selvarani.riic@gmail.com

R. Bharathi (✉)
Information Science and Engineering, PESIT-BSC, Bangalore, India
e-mail: rbharathi@pes.edu; sbharathi235@gmail.com

© Springer International Publishing AG 2017   247
A.K. Sangaiah et al. (eds.), *Intelligent Decision Support Systems*
for Sustainable Computing, Studies in Computational Intelligence 705,
DOI 10.1007/978-3-319-53153-3_13

This work focuses on modeling the failure analysis of SCSS using Universal Generating Function (UGF) [1], through the concept of Error Propagation (EP). The error propagation probability is a condition that once an error occurs in a system module, it might propagate to other modules and thereby cascades the error to the system output [2]. The error propagation analysis is a vital activity for the efficient and robust designing of safety critical software system. Generally, the functioning of SCSS is considered between two major states, perfect functioning and failure state. Here we are considering several intermittent states between the two major states for the failure analysis. Hence these systems can be termed as Multistate Systems (MS) in our research. The reliability of a MS can be defined as a measure of the capability of the system to execute required performance level [1].

The error in a software module might trigger an error across the other system modules that are interconnected [3]. Propagation analysis may be used to identify the critical modules in a system, and to explore how other modules are affected in the presence of errors. This concept will aid in system testing and debugging through generating required test cases, that will stimulate fault activation in the identified critical modules and facilitate error detection [4].

The kind of errors under consideration might be due to faulty design, which could result in errors and data errors due to wrong data, late data or early data. In this work, we assess the impact of EP across modules, by analyzing the error propagation process through probabilistic approach and arrive at a general expression to estimate the performance distribution of each module, subsystem and system. The reliability and performance of a multistate safety critical system can be computed by using Universal Generating Function (UGF) technique [5]. The UGF technique applied for analyzing the EP in safety critical systems in this paper is adapted by following the procedure demonstrated by Levitin et al. [5, 6].

The chapter is structured as follows: Related Terminologies are given in Sect. 2. Section 3 discusses on related works on error propagation. The influence of EP in software reliability prediction and universal generating function (UGF) are discussed in Sects. 4 and 5 respectively. The problem statement is explained in Sect. 6. Section 7 introduces the proposed approach through a framework. The error propagation and failure analysis of SCSS are discussed in Sect. 8. Section 9 gives the case study. Conclusion and future work are discussed in Sect. 10.

2 Terminologies

A *failure* is an event that occurs when the delivered service no longer complies with the expected service of the system. An *error* is an incorrect internal state that is liable to the occurrence of a failure or another error. However all errors may not reach the system's output to cause a failure. A *fault* is active when it results in an error otherwise it is said to be inactive [3]. Nevertheless, not all faults lead to an error, and not all errors lead to a failure. *Error propagation* (EP) can be defined as a condition where a failure of a component may potentially impact other system components or the environment through interactions, such as communication of bad data, no data, and early/late data [7].

3 Related Works

In the related area of our topic, there has been a wide discussion on software error propagation and reliability prediction by many authors. Jhumka et al. [8] derived a set metrics namely error transmission probability metric, error transparency metric and influence metric during the error propagation process. They identified vulnerable modules using error transmission probability and error transparency metrics. An analytical framework was developed to reduce the inter-modular Error Propagation in software.

In [9] the error propagation was analyzed and defined using probabilistic approach and derived two more concepts namely unconditional error propagation and cumulative error propagation among components. Finally they arrived with formulas to estimate the "error propagation probability", "unconditional and cumulative error propagation". The proposed formulas are validated through fault injection experiment.

The approach presented in [10], concludes that error propagation between the components have significant effect on the system reliability prediction. The propagation of errors between functions have been discussed in [11]. They have discussed on different approaches for error propagation assessment namely probabilistic approach, model based approach and formal approach. Further, they have defined a high level strategy on preventing error propagation and established a hypothesis.

A probabilistic analysis of error propagation due to either hardware or software faults in mechatronic systems are addressed in [12]. They have proposed an abstract mathematical framework called Error Propagation Model to analyze the error propagation in system level. Each system element is characterized by three parameters namely "Fault Activation", "Error Propagation" and "Error Detection" probabilities. They have used data flow graph to determine the error propagation between elements.

Morozov and Janschek [4, 13] have used probabilistic error propagation techniques for diagnosing the system. Henceforth it aids in tracing back the path of error propagation path to the error-origin. Moreover this diagnosis helps in error localization procedure, testing, and debugging. A bottom-up approach is considered to estimate the reliability for component-based system in [2]. Foremost, the reliability of system component was assessed. Based on the architectural information, the system reliability was estimated taking into the account of error propagation probability. The system analysis was carried out through the failure model by considering only data errors across components. Authors in [2] have concluded that error propagation is a significant characteristic of each system component and defined as the probability that a component propagates the erroneous inputs to the generated output. Their approach can be used in the early prediction of system reliability.

An error befalls when there is an activation of a fault [3]. An error occurs in a module when there is a fault in the module and henceforth it cannot directly cause an error in other modules. At many times, an error in a module can lead to its failure only within that module. The reason for the module error is either due to the activation of fault in the same module or due to deviated input service (failure) from other modules. A module failure is defined as the deviation of the performance from its accepted output behavior. If the failed module is the output interface module of the system then its failure is considered as a system failure [14]. System failures are defined based on its boundary. A system failure is said to occur when error propagates outside the system.

Filieri et al. [15] stated the all errors in the component are not liable to its failure. In turn, not all component failures necessarily lead to failure of the whole system. A component failure occurs only when an error propagates within the component up to its interface, and a system failure happens only when an error propagates across components up to the generated output. While during this propagation of error, there is a probability that an error can get masked, for example an erroneous value can be overwritten before being delivered to its interface. It is probable that an error can be transformed, from one type to another. For instance, a content failure received from one component could initiate additional computations, leading to timing failure [15]. In the propagation path each propagated error need not create the same kind of failures [16]. When a component fails, it is not that all failed components will propagate error and not all components will fail due to propagated error (error masking) [15]. Some systems are critical to certain category of failures and at the same time they might be less critical to other category of failures [17]. It is not necessary that the propagated error should create the similar kind of error in other components [18, 17].

To our knowledge, mostly researchers have discussed only on software error propagation. Embedded software is mostly used in safety critical systems to perform increasing number of safety critical functions. The fact is that these safety critical systems incorporate both software and hardware components with various mutual interactions. Hence, the occurrence of error in software due to failures caused by hardware components has to be accounted.

The error propagation analysis is an essential activity for the efficient and robust designing of safety critical software system.

In this paper, Safety Metric SM_{EP}, is considered, which has the following benefits [19]. It helps in assessing the probability of EP in the module level, subsystem and system level and to trace out the migration of error propagation from module level to subsystem level and to the system level.

4 Influence of EP in Software Reliability Prediction

Error Propagation has strong influence on system reliability. Many researchers [2, 10, 17, 18, 20, 21] have detailed the significance of EP inclusion in the architecture based reliability assessment. They have noted that many existing reliability models are based on black box approach. Moreover each researcher has different viewpoints on error propagation and characterized in several ways.

Identification of different error states and failure modes of each component or module is very essential. Each error type may have different propagation path. Probabilistic error propagation analysis is the most preferred method by many researchers. Fiondella and Gokhale [21] presented an approach to find the reliability of the software at architectural design stage through EP modeling. To achieve this each component is characterized with six parameters namely probability of correct output for a given correct input, probability of incorrect output for a given correct input, probability of no output for a given correct input, probability of correct output for a given incorrect input, probability of incorrect output for a given incorrect input, probability of no output for a given incorrect input. Through which the criticality of the components are identified such that it can be equipped with error recovery mechanisms to improve the system reliability.

Reliability at its extensive level is considered as performance measure and stated as the capability of an entity to perform a required function under specified condition for a specified period of time.

5 Universal Generating Function Technique for MS

The UGF technique is based on probability theory to assess and express models through polynomial functions. It is also called as u-function introduced by Ushakov [22] and Levitin [5] expanded and proved that UGF is an effective technique for assessing the performance of real world systems, in specific Multistate Systems. In general all traditional reliability models perceived a system as binary state systems, states being perfect functionality and complete failure. In reality, each system has different performance levels and various failure modes affecting the system performance [1]. Such systems are termed as Multistate-Systems (MS).

Let us assume an MS composed of n modules. In order to assess the reliability of a MS, it is necessary to analyze the characteristic of each module present in the system. A system module 'm' can have different performance rates and represented by a finite set q_m, such that $q_m = \{q_{m1}, q_{m2}, \ldots \ldots q_{mi} \ldots q_{mk_m}\}$ [23], where q_{mi} is the performance rate of module m in the ith state and $q_i = \{1, 2, \ldots k_m\}$. The performance rate $Q_m(t)$ of module 'm', at time t $\geq$ 0 is a random variable that takes its value from $q_m: Q_m(t) \in q_m$.

Let the ordered set $p_m = \{p_{m1}, p_{m2}, \ldots p_{mi}, \ldots p_{mj_m}\}$ associate the probability of each state with performance rate of the system module m, where $p_{mi} = Pr\{Q_m = q_{mi}\}$. The mapping $q_{mi} \rightarrow p_{mi}$ is called the probability mass function (pmf) [24].

The random performance [25] of each module m is defined as polynomials can be termed as module's UGF denoted as "$u_m(z)$",

$$u_m(z) = \sum_{i=0}^{k} P_{mi} z^{q_{mi}}, m = 1, 2 \ldots n. \tag{1}$$

Similarly the performance rates of all 'm' system modules have to be determined. At each instant $t \geq 0$, all the system modules have their performance rates corresponding to their states. The UGF for the MS denoted as "$(U_S(z))$" can be arrived, by the determining the modules interconnection through system architecture. The random performance of the system as a whole at an instant $t \geq 0$ dependent on the performance state of its modules. The UGF technique specifies an algebraic procedure to calculate the performance distribution of the entire MS, denoted as $U_S(z)$, let $U_s(z) = f\{u_{m1}(z), u_{m2}(z), \ldots, u_{mn}(z)\}$,

$$U_s(z) = \nabla_\phi \{u_{m1}(z), u_{m2}(z), \ldots, u_{mn}(z)\} \tag{2}$$

where, ∇ is the composition operator and ϕ is the system structure function. In order to assess the performance distribution of the complete system with the arbitrary structure function ϕ, a composition operator ∇ is used across individual u function of m system modules [24].

$U_S(z)$, is U-function representation of performance distribution of the whole MS software system. The composition operator ∇ determines the U function of the whole system by exercising numerical operations on the individual u functions of the system modules. The structure function $\phi(\cdot)$ in composition operator ∇ expresses the complete performance rate of the system consisting of different modules in terms of individual performance rates of modules. The structure function $\phi(\cdot)$ depends upon the system architecture and nature of interaction among system modules.

Reliability is nothing but continuity of expected service [3] and it is well known that, it can be quantitatively measured as failures over time. The UGF technique can be used for estimating of software reliability of the system as a whole consisting of n module. Each of the modules performs a sub function and the combined execution of all modules performs a major function [24]. An assumption while using the UGF technique is that the system modules are mutually independent of their performance.

In MS, the reliability (R_{MS}) at instant 't' can be defined as the probability that a system as whole can operate and perform the required service "S". Hence the performance rate of the MS at instant 't' can be represented, as 'B' and that should be greater than or equal to "S".

$$R_{MS}(t, S) = \Pr\{B \geq S\} \tag{3}$$

To assess the reliability of the MS system, we need to estimate the performance of the system as whole as shown in Eq. 2.

6 Problem Statement

Error Propagation (EP) is defined as the condition where there is a probability of an error (or failure) propagates across various modules or components in the SCSS [26]. Our approach focuses on quantifying the propagation of error between modules in safety critical software system using UGF, because it provides a practical adaptation concept to facilitate appropriate actions in complex and changing environments [1].

7 Proposed Approach

The analysis proposed in this research is explained through a framework as shown in Fig. 1. To begin, the performance distribution of system modules called PD_{MOD} is determined using u-function.

The probability of error propagation at module level ($PD_{MOD} + SM_{EP}$) is quantified in the second step. As third step, the performance distribution of subsystems is arrived through composition operator ∇ having a structure function $\phi(\cdot)$. As the final step the failure prediction is achieved through recursive operations for quantifying the error propagation throughout the system. During software development, this framework would be helpful to demonstrate the probability of error propagation to identify the error prone areas. The estimated performance of the system helps to assess the reliability.

8 EP and Failure Analysis

The EP and failure analysis model is a conceptual framework for analyzing the occurrence of error propagation in SCSS [19]. The system considered is broken down into subsystem, and each subsystem in turn is subdivided into modules called elements. A module is an atomic structure, which performs definite function(s) of a complex system.

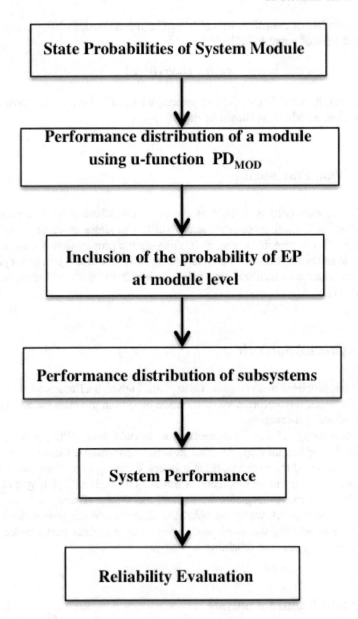

Fig. 1 Framework for EP and failure analysis

8.1 Performance Distribution of a Module

The performance rate of a module can be measured in terms of levels of failure [19].

Let us assume that the performance rate of a module m with 0% failure is q_{m1}, 10% failure is q_{m2}, 30% failure is q_{m3}, 50% failure is q_{m4} and 100% failure is q_{m5}.

The state of each module m can be represented by a discrete random variable Q_m that takes value form the set,

$$Q_m = \{q_{m1}, q_{m2}, q_{m3}, q_{m4}, q_{m5}\} \tag{4}$$

The random performance of a module varies from perfect functioning state to complete failure state.

The probabilities associated with different states (performance rates) of a module m at time t can be represented by the set,

$$P_m = \{p_{m1}, p_{m2}, p_{m3}, p_{m4}, p_{m5}\}, \text{where}, P_{mh} = \Pr\{Q_m = q_{mh}\}.$$

The module's states is the composition of the group of mutually exclusive events,

$$\sum_{h=1}^{5} P_{mh} = 1 \tag{5}$$

The performance distribution of a module m (pmf of discrete random variable G) can be defined as

$$u_m(z) = \sum_{h=1}^{5} P_{mh} z^{q_{mh}} \tag{6}$$

The performance distribution of any pair of system modules l and m, connected in series or parallel [25] can be determined by,

$$u_l(z) \nabla u_m(z) = \sum_{h=1}^{5} P_{lh} z^{q_{ih}} \nabla \sum_{h=1}^{5} P_{mh} z^{q_{mh}} \tag{7}$$

The composition operator ∇ determines the u function for two modules based on whether they are connected in parallel or series using the structure function ø. The equation arrived in Eq. (7) quantifies the performance distribution of combination of modules. Levitin et al. in [6] have demonstrated the determination of performance distribution when the modules are connected in series or parallel.

8.2   Safety Metric SM_{EP}

The probability of occurrence of EP in a module can be defined by introducing a new state in that module [6]. Assuming that the state 0 of each module corresponds to the EP originated from this module [5]. The Eq. (6) can be rewritten as,

$$u_m(z)_{ep} = P_{m0}z^{q_{m0}} + \sum_{h=1}^{5} P_{mh}z^{q_{mh}} \tag{8}$$

$$u_m(z)_{ep} = P_{m0}z^{q_{m0}} + u_m(z) \tag{9}$$

where p_{m0} is the probability state for error propagation and q_{m0} is the performance of the module at state 0. $u_m(z)$ represents all states except the state of error propagation.

The performance distribution of a module m at state 0 is the state of error propagation is given by $P_{m0}z^{q_{m0}}$ is termed as Safety Metric SM_{EP}, used to measure the probability of error propagation. The Safety metric SM_{EP} [19] of each module will carry a weightage based on the probability of propagating error. If the module does not propagate any error, the corresponding state probability should be equated to zero [6].

$$p_{m0} = 0 \tag{10}$$

By substituting Eq. (10) in Eq. (8), the SM_{EP} is quantified as zero. Therefore Eq. (8) becomes,

$$u_m(z)_{ep} = \sum_{h=1}^{5} P_{mh}z^{q_{mh}} \tag{11}$$

If the module that does not have error propagation property or state then Eq. (11) will be reduced to Eq. (6).

If the module can cause error propagation, then the performance of the module in that state of error propagation is

$$q_{m0} = \alpha \tag{12}$$

The value of α in Eq. (12) can be any random performance q_{m1} or q_{m2} or q_{m3} or q_{m4} or q_{m5}. The conditional pmf of any operational module m that will not fail due to error propagation can be represented by u-function [6],

$$u_m(z)_{ep} = \sum_{h=1}^{5} \frac{p_{mh}}{1 - p_{m0}} z^{q_{mh}} \tag{13}$$

Because the module can be in any one of the five states as defined in Eq. (6). The Safety metric SM_{EP} depends on the performance of each module in the multistate system. As per [19], this safety metric helps to measure whether the module or the subsystem has the influence of EP or not. Hence the performance distribution of module, subsystem and system is dependent on this safety metric.

8.3 Module Definition in Terms of PD_{MOD} and SM_{EP}

The safety metric SM_{EP} depends upon the performance of each module. This is the second step of our EP and failure analysis framework as depicted in the Fig. 2. The safety metric SM_{EP} of each module will carry a weightage based on the level of interaction with other modules of the system and the impact of error propagation to the other modules and within itself.

SM_{EP} = Function (Performance of module w.r.t propagation of error)

In this aspect, each module called MOD, can be defined by the following tuple,

MOD = <PD_{MOD}, SM_{EP}>
PD_{MOD} = Performance distribution of module in terms of levels of failure
SM_{EP} = Module Safety Metric.

The estimation of safety metric SM_{EP} depends upon the probability of occurrence of EP among modules in the system as described in the Eqs. (9–13).

8.4 Performance Distribution of Subsystem PD_{SS}

The subsystem performance depends upon the performance of all modules present in the subsystem. Because there is probability that migration of error occurs from

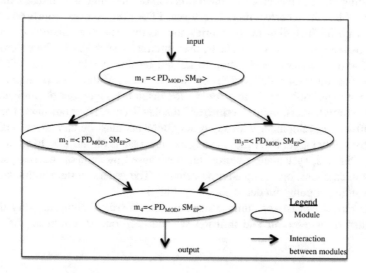

Fig. 2 An example subsystem

modular level to subsystem level. Hence the performance of subsystem depends on the module performance $PD_{MOD,}$ module safety metric SM_{EP} and subsystem structure function $\emptyset(\cdot)$. The function $\emptyset(\cdot)$ depends upon the nature of modules interaction in the subsystem. Dependency Graph (DG) is established method for conveying the architectural dependency between modules [27] and indicating the possible execution of modules. An example subsystem is depicted in the Fig. 2.

Each subsystem is defined by a quadruple $<N_m, \emptyset(\cdot), PD_{SS}, SM_{EP}>$,

where,

N_m Number of modules in the subsystem
$\emptyset(\cdot)$ Structure function determines the type of connection and nature of interaction among modules in the subsystem
PD_{SS} Performance distribution of subsystem in terms of levels of failure
SM_{EP} Safety Metric.

Depending upon the subsystem architecture, the u function of each subsystem can be quantified by applying the composition operator $\nabla_\emptyset$. In a subsystem, if a failed module is the output interface module, then its failure is considered as subsystem failure. Hence there will be a probability of error propagation, outside this subsystem. Then the recursive approach is used to obtain the entire u-function of safety critical software system.

9   Case Study

To illustrate our proposed approach on EP and failure analysis, we have taken Insulin Infusion Pump (IIP) as a case study. IIP is software intensive medical device for treating diabetic patients. The main function of this pump is to infuse calculated amount of insulin at correct time. Every year FDA [28] receives reports on adverse events with IIP including many injuries and deaths. The most frequently received problems are due to software defects, user interface issues and hardware problems [28]. The most hazardous situation in IIP are insulin overdose and insulin under dose which are insulin delivery errors, which occurs at the system output. In all modern pumps, software is responsible for insulin dosage control, mitigation of hazards through alerts, input interface and display. Hence, laborious hazard analysis and software development with safety requirements must be carried out and validated for safety usage. FDA [28] have listed out reasons for hazards, to name a few, random failures, systematic failures, failures caused by component, subsystem or system interactions, operating environment etc. The generic system architecture for insulin infusion pump model is shown in Fig. 3.

This model consists of pump controller, user interface, pump delivery mechanism, and drug reservoir and infusion set and each one is described with stated

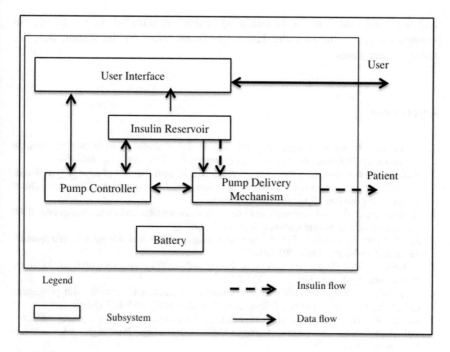

Fig. 3 Generic block diagram of insulin infusion pump

scope. The "Pump controller" is a subsystem responsible for "Computation" for the insulin infusion pump as a whole. The pump controller computes the dose of insulin for administrating the patient. Four types of insulin can be administered to a patient, namely basal insulin, temporary insulin, bolus and extended bolus insulin. In order to determine the IIP is adequately safe, a rigorous failure analysis has to be conducted. The performance distribution of the pump can be assessed using our proposed safety metric SM_{EP}. Numerical calculations to assess the performance of the pump through our EP will be discussed in our subsequent work. An elaborate exercise has to be carried on all possible failure behavior of the pump, to locate the occurrence of error and subsequently the error propagation.

10   Conclusion and Future Work

The EP and Failure Analysis framework helps to analyze the failure of multistate safety critical software. The proposed metric SM_{EP} has the application in finding the failure probability of each module, the migration of error propagation from modular level to subsystem and then to system level. Subsequently it helps in the process of identifying the most critical module across the safety critical software system and the impact of error propagation in the performance of SCSS. Hence this method provides the base for the reliability evaluation, since the occurrence of error propagation across the modules has a significant effect on the system behavior during critical states.

References

1. G. Levitin, A universal generating function in the analysis of multi-state systems, in *Handbook of Performability Engineering* (Springer, London, 2008), pp. 447–464
2. V. Cortellessa, V. Grassi, A modeling approach to analyze the impact of error propagation on reliability of component-based systems, in *International Symposium on Component-Based Software Engineering* (Springer, Berlin, 2007)
3. A. Avizienis et al., Basic concepts and taxonomy of dependable and secure computing. IEEE Trans. Dependable Secure Comput. **1**(1), 11–33 (2004)
4. A. Morozov, K. Janschek, Probabilistic error propagation model for mechatronic systems. Mechatronics **24**(8), 1189–1202 (2014)
5. G. Levitin, Block diagram method for analyzing multi-state systems with uncovered failures. Reliab. Eng. Syst. Saf. **92**(6), 727–734 (2007)
6. G. Levitin, L. Xing, Reliability and performance of multi-state systems with propagated failures having selective effect. Reliab. Eng. Syst. Saf. **95**(6), 655–661 (2010)
7. P.H. Feiler, J.B. Goodenough, A. Gurfinkel, C.B. Weinstock, L. Wrage, Reliability validation and improvement framework. Carnegie-Mellon University, Pittsburgh, PA, Software Engineering Institute, 2012
8. A. Jhumka, M. Hiller, N. Suri, Assessing inter-modular error propagation in distributed software, in *Proceedings. 20th IEEE Symposium on Reliable Distributed Systems, 2001* (IEEE, 2001)
9. D.E. Nassar, W.A. Rabie, M. Shereshevsky, N. Gradetsky, H.H. Ammar, S. Bogazzi, A. Mili, Estimating error propagation probabilities in software architectures 1 (2004)
10. P. Popic et al., Error propagation in the reliability analysis of component based systems, in *16th IEEE International Symposium on Software Reliability Engineering (ISSRE'05)* (IEEE, 2005)
11. R. Fredriksen, R. Winther, Challenges related to error propagation in software systems, in *Proceedings of Risk, Reliability and Societal Safety (ESREL)*, 2007, pp. 83–90
12. A. Morozov, K. Janschek, Case study results for probabilistic error propagation analysis of a mechatronic system. TagungsbandFachtagungMechatronik 229–234 (2013)
13. A. Morozov, K. Janschek, Dual graph error propagation model for mechatronic system analysis. IFAC Proc. **44**(1), 9893–9898 (2011)
14. A. Mohamed, M. Zulkernine, Failure type-aware reliability assessment with component failure dependency, in *2010 Fourth International Conference on Secure Software Integration and Reliability Improvement (SSIRI)* (IEEE, 2010)
15. A. Filieri et al., Reliability analysis of component-based systems with multiple failure modes, in *Component-Based Software Engineering* (Springer, Berlin, 2010), pp. 1–20
16. N.A.M. Alzahrani, D.C. Petriu, Modeling component erroneous behavior and error propagation for dependability analysis, in *SDL 2013: Model-Driven Dependability Engineering* (Springer, Berlin, 2013), pp. 124–143
17. A. Mohamed, M. Zulkernine, Architectural design decisions for achieving reliable software systems, in *Architecting Critical Systems* (Springer, Berlin, 2010), pp. 19–32
18. A. Mohamed, M. Zulkernine, On failure propagation in component-based software systems, in *The Eighth International Conference on Quality Software, 2008. QSIC'08* (IEEE, 2008)

19. R. Selvarani, R. Bharathi, A novel safety metric SM_{EP} for performance distribution analysis in software system. in *Strategic Engineering for Cloud Computing and Big Data Analytics*, Springer, 2017 (in press).
20. T.-T. Pham, X. Défago, Q.-T. Huynh, Reliability prediction for component-based software systems: dealing with concurrent and propagating errors. Sci. Comput. Program. **97**, 426–457 (2015)
21. L. Fiondella, S.S. Gokhale, Architecture-based software reliability with error propagation and recovery, in *2013 International Symposium on Performance Evaluation of Computer and Telecommunication Systems (SPECTS)* (IEEE, 2013)
22. I.A. Ushakov, A universal generating function. Soviet J. Comput. Syst. Sci. **24**(5), 118–129 (1986)
23. G. Levitin, A. Lisnianski, Multi-state system reliability: assessment, optimization and applications (2003)
24. G. Levitin, *The Universal Generating Function in Reliability Analysis and Optimization* (Springer, London, 2005)
25. G. Levitin, T. Zhang, M. Xie, State probability of a series-parallel repairable system with two-types of failure states. Int. J. Syst. Sci. **37**(14), 1011–1020 (2006)
26. S. Sarshar, Analysis of error propagation between software processes. Nucl. Power-Syst. Simul. Oper. 69 (2011)
27. S. Yacoub, B. Cukic, H.H. Ammar. A scenario-based reliability analysis approach for component-based software. IEEE Trans. Reliab. **53**(4), 465–480 (2004)
28. http://www.fda.gov/downloads/Training/CourseMaterialsforEducators/NationalMedicalDev-iceCurriculum/UCM404248.pdf

A Framework for Analyzing Uncertainty in Data Using Computational Intelligence Techniques

M. Sujatha, G. Lavanya Devi and N. Naresh

Abstract In this study, use of soft computing techniques for analyzing medical data is presented. A medical dataset usually contains objects/records of patients that include a set of symptoms that a patient experiences. Analysis of such medical data could reveal new insights that would definitely help in efficient diagnosis and also in drug discovery. A novel fuzzy-rough based classification approach is described and its performance is evaluated using a medical dataset having multiclass values for response variable. Novel approaches for data preprocessing using fuzzy-rough concepts are introduced for attaining complete and consistent data. Thus, automated medical diagnosis can be done efficiently by using computational intelligence (CI) techniques for the benefit of mankind to live a healthy long life.

Keywords Rough set theory · Fuzzy set theory · Fuzzy-rough concepts · Computational intelligence

1 Introduction

Artificial Intelligence (AI) is an emulating human intelligence on machines to act and think similar to human beings. It contains reasoning, planning, intelligent search, machine learning and perception building. Traditional AI problem resolving methods are focused on problem states and rule set design to draw transitions in problem states. AI techniques are better for inductive and analogy-based learning compared to supervised learning [1]. These techniques are less feasible for opti-

M. Sujatha · G. Lavanya Devi (✉) · N. Naresh
Computer Science and System Engineering,
Andhra University College of Engineering, Visakhapatnam, India
e-mail: lavanyadevig@yahoo.co.in

M. Sujatha
e-mail: sujathamadugulacse@gmail.com

N. Naresh
e-mail: naresh855@gmail.com

© Springer International Publishing AG 2017
A.K. Sangaiah et al. (eds.), *Intelligent Decision Support Systems for Sustainable Computing*, Studies in Computational Intelligence 705, DOI 10.1007/978-3-319-53153-3_14

mization and machine learning with regard to uncertain data. These failures of traditional AI opened up with solutions from CI techniques for real world problems. CI adopts techniques motivated by nature, that possess the ability to learn and deal with new situations with high computational speed. Also, they are less error-prone to noisy information sources [2].

1.1 Motivation

There are several data mining techniques to analyze data. Still there is always a scope for new approaches to analyze data for better decision making. The current trend is to develop techniques for data analysis based on domain, volume and type of the data. This approach has every chance of better decision making than generic approaches. Hence, techniques tailored based on the domain (banking, finance, social media, medical etc.) would perform better. Every domain has some or other form of uncertainty that has to be interpreted properly. This research is focused on medical domain.

In general, medical data has uncertainty due to the reason that a patient suffering from a specific illness cannot be completely determined by one or more symptoms; a certain set of symptoms can only indicate that there is a probability of a particular illness.

Broadly medical misdiagnosis can be classified into three classes.

False positive: misdiagnosis of a disease that is not actually present.
False negative: failure to diagnose a disease that is present.
Equivocal results: inconclusive interpretation without a definite diagnosis [60].

General data mining methods are inefficient in dealing with cognitive uncertainties such as vagueness and ambiguity.

Proper analysis of medical datasets is one of the best ways to discover potentially useful information for diagnosis and also drug discovery. This research work aims to address the problem of uncertainty in medical datasets. Fuzzy-rough set based data preprocessing methodologies have been proposed for handling medical datasets. Finally an ensemble of rule based fuzzy-rough classifier is built for analyzing medical datasets.

The proposed work has been carried out to achieve the following objectives that address the above mentioned problem.

1.2 Objectives of the Proposed Work

To study the role of uncertainty in real world problems and to understand the scope of RST for analyzing uncertain data.

To design an efficient classifier that handles ambiguity and vagueness in medical datasets for better diagnosis of illness.

To design and built a rule based fuzzy-rough classifier for analyzing uncertainty in medical dataset.

To perform experimental analysis to evaluate the performance of the proposed classification model.

To test the performance of the proposed approach with existing state-of-the-art approaches.

1.3 Computational Models for Prediction

In general, systems built based on CI techniques are used for decision making. Decision making is essential in all fields of human activities and affect every sphere of our life. Today's real world problems are mostly data driven. Data accumulation in every domain of life is a current challenge for data engineers. Often, data is imprecise, ambiguous, vague and uncertain. Decision making is a difficult process due to the uncertainty in data. Firstly, data has to be set free from ambiguity and uncertainty. Then, it can be used for decision making. These situations can be better handled with CI based solutions. CI techniques constitute of neural networks (NN), genetic algorithms (GA), fuzzy systems, rough set and hybrid systems (combinations of NN, GA, fuzzy system and rough sets).

Neural networks learn by instances. Thus, NN architectures identifies known instances of a problem before they are tested for their 'inference' capability on known instances of the problem [3]. Therefore, it identifies new instances before untrained. It has ability to generalize and predict new outcomes.

Genetic algorithms are computerized search and optimization techniques based on procedure of selecting natural genetics. GA and evolutionary strategies mimic the principle of natural genetics and natural selection to construct search and optimization procedures [4].

Fuzzy set theory is proven to solve uncertainty in data. Rough set is a new approximation of a crisp set. It deals with uncertainty in data by using approximation [5]. Uncertain data in a given model can be handled by indiscernibility for rough sets and vagueness for fuzzy sets.

2 Uncertainty in Data

Uncertainty may be due to lack of knowledge or insufficient information [6–9]. Vagueness and ambiguity are the two major forms of uncertainty. Vagueness is associated with the difficulty of making sharp or precise distinctions in the real world [10, 11].

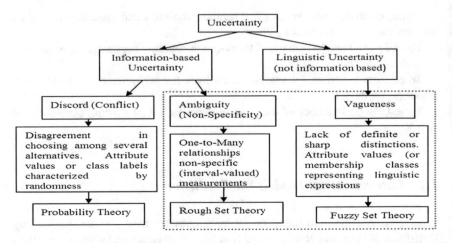

Fig. 1 Classification of uncertainty

Ambiguity is associated with two or more alternatives such that the choice between them is left unspecified. Broad classification of uncertainty is depicted in Fig. 1.

Information based uncertainty arises from lack of information. It is classified into discord and ambiguity. Discord is associated with the conflict in choosing among several alternatives of the attribute. It is handled by the probability theory [12]. Ambiguity can be addressed by rough set theory, to discover hidden patterns in data. It finds partial or total dependencies in databases, eliminates redundant and missing data.

Linguistic uncertainty that arises in natural language is vague and also the precise meaning of the words can change over time. Linguistic variables are words or sentences in a natural or artificial language. The meaning of linguistic variable is given as vagueness. Vagueness is existence of objects which cannot be uniquely classified relative to a set or its complement [1].

2.1 Classical Set Theory Versus Fuzzy Set Theory Versus Rough Set Theory

This section provides an overview of classical set, fuzzy set and rough set theories which includes the basic concepts, notations, applications and limitations. A summary of the same is given in Table 1.

Classical Set Theory

A set is defined as collections of objects which share certain characteristics. Set theory is the branch of mathematical logic that studies sets. It was initiated by

Table 1 Classical set theory versus fuzzy set theory versus rough set theory

	Classical set theory	Fuzzy set theory	Rough set theory
Definitions	Elements can belong to a set or not at all. It either belongs to the set definitely (denoted as 1) or does not belong to the set definitely (denoted as 0). Also known as crisp sets	Fuzzy sets are those whose elements have degrees of membership. It permits the gradual assessment of the membership of elements valued in the real unit interval [0, 1]. Crisp sets are special cases of fuzzy sets	Rough set theory is an extension to set theory that enables approximation in decision making by using a pair of sets {*Lower Approximation, Upper Approximation*} of original set. Rough set is a formal approximation of a crisp set
Notations	Let U be a space of objects, referred to as the universe of discourse, and x an element of U. A classical (crisp) set A, $A \subseteq U$ is defined as a collection of elements $x \in U$, such that each element x can belong to the set or not belong	Let U be a finite and non-empty set called universe. A fuzzy set A of U is defined by a membership function: $\mu_A: U \to [0,1]$	The decision system $(U, A \cup \{d\})$, where U is the finite universe of discourse, A the set of conditional attributes and d the decision attribute, with $d \notin A$. $P \subseteq A$ there is associated an equivalence relation $IND_A(P)$: $\{(x, y \in U^2) \vert \forall_a \in P, a(x) = a(y))\}$
Operations	*Union:* $A \cup B = \{x \vert (x \in A) \, or \, (x \in B)\}$ *Intersection:* $A \cap B = \{x \vert (x \in A) \, and \, (x \in B)\}$ *Difference:* $B - A = \{x \vert (x \in B) \, and \, not \, (x \in A)\}$ *Complement:* $A^c = \{x \vert (x \notin A), (x \in U)\}$ where A and B are crisp sets	*Union:* $\mu_{A \cup B}(x) = max(\mu_A(x), B(x))$ *Intersection:* $\mu_{A \cap B}(x) = min(\mu_A(x), B(x))$ *Difference:* $A - B = (A \cap \bar{B})$ *Complement:* $\mu_A(x) = 1 - \mu_A(x)$ where A and B are fuzzy sets	*Lower approximation:* $\underline{P}X = \bigcup_{x \in U} \{P(x): P(x) \subseteq X\}$ *Upper approximation:* $\overline{P}X = \bigcup_{x \in U} \{P(x): P(x) \cap X \neq \emptyset\}$ *Positive region:* $Pos_P(Q) = \bigcup_{X \in U/Q} \underline{P}X$ *Boundary region:* $BND_P(Q) = \bigcup_{X \in U/Q} \overline{P}X - \bigcup_{X \in U/Q} \underline{P}X$
Applications	Used in different mathematical approaches like vector spaces, graphs, rings and manifolds described by axiomatic properties	Used for knowledge discovery in databases, to design controllers for electronic instruments, automobiles and traffic monitoring systems	Used in computational intelligence, pattern recognition, acoustic analysis and metrological pattern classification
Limitations	No partial membership is allowed	Membership value of an element does not depend on other elements	Membership value of an element depends on other elements

Georg Cantor and Richard Dedekind in 1870s. After the discovery of paradoxes in naive set theory, numerous axiom systems were proposed in the early twentieth century, of which the Zermelo–Fraenkel axioms, with the axiom of choice, are the best-known [47]. Table 1 gives a summary on the basics, operations and applications of set theory.

Fuzzy Set Theory

Fuzzy Set Theory permits gradual assessment of membership of elements in a set with real value interval [0, 1]. It represents classical bivalent sets as *crisp* sets [13]. It is used when information is vague, incomplete or imprecise. In real life, human imprecision causes an information available to be vague or fuzzy. Vagueness is handled by making use of "soft" boundaries of fuzzy sets i.e., graded membership which gives subjective knowledge to define these attributes. Fuzzifying attributes is tedious when precise information is available. Brief on basics, operations and applications of fuzzy theory is depicted in Table 1.

Rough Set Theory

Rough Set Theory is an extension of conventional set theory that supports approximations in analyzing decisions [14]. It is represented as a pair of sets, *Lower* and *Upper* approximation of crisp set. Lower approximation identifies objects that certainly belong to subset of interest, whereas upper approximation identifies objects that possibly belong to subset [15–18].

Rough set theory based models efficiently handles incomplete or imperfect knowledge [19–21]. Table 1 gives the outline of basics, operations and applications of classical set, fuzzy set and rough set theories.

2.2 Combining Fuzzy and Rough Set Theories

Fuzzy and rough set theories evolved as successful approaches to represent and compute imperfect data in real world applications [22–24]. Both of the theories are not competitive rather they are complementary to each other [25–27]. Along with the complementary nature, the similarities between them have motivated to develop a hybrid theory that covers the strengths of each. Lynn Deer et al., have proposed the process of fuzzifying the lower and upper approximations [28–30].

In this theory, given incomplete information (A) representing a subset of given universe (U) containing examples of a concept (C), the lower and upper approximations, along with an equivalence relation (R) that models indiscernibility can be expanded in the following two ways:

i. Objects in set A can belong to a concept with varying degree, i.e., making the set into a fuzzy set.

ii. Objects are classified into classes with soft boundaries, using similarities among the objects which are represented by a fuzzy relation R. Here, the fuzzy relation is used instead of indiscernibility relation.

This article does not intend to cover fuzzy-RST in its entirety. Rather, it confines to a brief introduction of rough-fuzzy and fuzzy-rough concepts.

Rough-Fuzzy Sets

A rough-fuzzy set is simplified from approximation of a fuzzy set in a crisp approximation space, where decision attribute values are fuzzy and conditional values are crisp [42]. In rough-fuzzy set, lower and upper approximations of objects belonging to these sets are defined as

$$\mu_{\underline{P}X}\left([x]_P\right) = inf\left\{\mu_X(x)|x \in [x]_P\right\} \tag{1}$$

$$\mu_{\overline{P}X}\left([x]_P\right) = sup\left\{\mu_X(x)|x \in [x]_P\right\} \tag{2}$$

where $[x]_P$ is crisp, $\mu_X(x)$ is degree of x fit into fuzzy equivalence class X, tuple $<\underline{P}X, \overline{P}X>$ is represented as rough-fuzzy set. Rough-fuzzy sets simplified to fuzzy-rough sets, when equivalence classes are fuzzy.

Fuzzy-Rough Sets

Competing with rough-fuzzy set theory, researchers focused on fuzzy-rough set theory [44, 45]. Earlier, such type of hybridization makes data analysis on information systems. Dubois suggested fuzzy-rough set theory which approximate fuzzy sets [43]. Rough sets are defined in terms of fuzzy membership function. It represents boundary regions, positive and negative region. In boundary region objects belong to membership value is 0.5, positive region objects belong to membership value is 1 and negative region objects belong to membership value is 0.

2.3 Neural Networks

Neural networks are simplified models of the biological nervous systems. They can be defined as data processing systems, consisting of a large number of simple, highly interconnected processing elements (artificial neurons), in an architecture inspired by the structure of the cerebral cortex of the brain.

Learning methods in NN are broadly classified into three types: *supervised, unsupervised and reinforced* [31].

In supervised learning, every input pattern that is used to train the network is associated with an output pattern, which is the target or the desired pattern.

In unsupervised learning, the target output is not presented to the network.

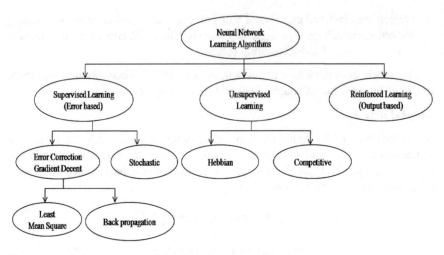

Fig. 2 Classification of learning algorithms

Table 2 The classification of NN system with respect to type of architecture

Type of architecture			
	Single layer feedforward network	Multilayer feedforward network	Recurrent network
Diagram			
Description	Comprises of two layers: input layer and output layer	Comprises of multiple layers: input layer, hidden layer (one or more) and output layer	Comprises of multiple layers as in feedforward networks. In addition, they have at least one feedback loop.
Popular methods	ADALINE, Hopfield, Perceptron	Cascade correlation, radial basis function	Adaptive resonance theory, Boltzman machine, Cauchy machine
Remarks	Can only deal linear problems	Can deal non-linear problems	It is often difficult to predict the optimal neural network size for a particular application
		With the increase of number of hidden layers computational complexity increases	

In reinforced learning, indication whether the computed output is correct or incorrect is given. A reward is given for the correct output and incorrect output is penalized. Classification of learning algorithms as depicted in Fig. 2.

Table 2 shows the classification of the NN systems listed above, according to their learning methods and architectural types.

2.4 Genetic Algorithms

Genetic algorithms are adaptive heuristic search methods based on principles of evolutionary ideas of natural selection and genetics [40]. GAs represents decision attributes of a search problem into fixed-length strings of certain cardinality. Strings results as candidate solutions to search problem for specified chromosomes, alphabets are referred to as genes and values of genes are named as alleles. When problem is encoded in a chromosomal manner and a fitness measure for discriminating good solutions from bad ones have been chosen, solutions to search problem start to evolve using *Initialization, Evaluation, Selection, Recombination and Mutation* [41].

Initial population of candidate solutions is generated randomly across the search space. When population is initialized, fitness values of the candidate solutions are evaluated. Selection allocates more copies of those solutions with higher fitness values and thus imposes the survival-of-the-fittest procedure on candidate solutions. Recombination combines parts of two or more parental solutions to create new, possibly better solutions. The offspring under recombination will not be identical to any particular parent. Table 3 presents most commonly used selection methods, recombination (crossover) operators, mutation operators and replacement of GAs [39].

While recombination operates on two or more parental chromosomes, mutations are changes in genetic sequence of a chromosome. These changes occur at many different levels of individual chromosomes. The offspring population created by selection, recombination, and mutation replaces the original parental population.

2.5 Hybrid Systems

Hybrid systems is combination of two or more techniques to overcome limitations of individual techniques. It is impossible to deal with either as purely continuous or discrete-event system without ignoring important phenomena that result from the combination of continuous and discrete movements of this system [35].

Mohammed Hamed Ahmed Elhebir described minimum support requirement dictates the efficiency of association rule mining [48]. If support threshold is low, then not truly interesting rules are generated. On the other hand, if the support threshold is high, then interesting rules are missed from the rule set. Eliminating redundant rules and clustering decreased the size of the generated rule set for obtain interestingness rules.

Abraham et al., presented biological motivation on particle swarm optimization and ant colony optimization algorithms. The basic data mining terminologies are explained by using swarm intelligence techniques [49].

Table 3 Genetic algorithm operators (GAO)

GAO	Methods	Description
Selection methods	Fitness proportionate selection	Selection is performed based on the fitness function derived
		Methods mostly used are Roulette-wheel Selection, Boltzmann Selection, Rank Selection, Steady State Selection and Stochastic Universal Selection
	Ordinal selection	Selection procedure is independent of fitness function
		Methods mostly used are Tournament Selection and Truncation Selection
Recombination (crossover) operators	Crossover operators	Individuals from the mating pool are recombined to create new, hopefully better, offspring
	k-point crossover	Crossover site is selected at random over string length and alleles on one side of site are exchanged between individuals
	Uniform crossover	Every allele is exchanged between a pair of randomly selected chromosomes with a certain probability
	Uniform order-based crossover	k-point and uniform crossover methods described above are not well suited for search problems with permutation codes such as the ones used in the traveling salesman problem
	Order-based crossover	It is a variation of uniform order-based crossover in which two parents are randomly selected and two random crossover sites are generated
	Partially matched crossover	Two parents are randomly selected and two random crossover sites are generated
	Cycle crossover	Two randomly selected parents P_1 and P_2 that are solutions to a traveling salesman problem. Offspring C_1 receives first variable from P_1. Then choose variable that maps onto same position in P_2
Mutation operators	It is designed if two parents has same allele at a given gene then one-point crossover will not change that i.e., gene have same allele forever	
Replacement	Delete-all	Deletes all members of current population and replaces them with same number of chromosomes that have just been created
	Steady-state	Deletes n old members and replaces them with n new members. The number to delete and replace, n, at any one time is a parameter to this deletion technique
	Steady-state-no-duplicates	It checks no duplicate chromosomes are added to population

Table 4 Hybrid systems and their applications

Hybrid system	Applications
KMLP-ANN (K-means Multilayer Perceptron Artificial Neural Networks)	It investigates k-means algorithm for data filtering and MLP-ANN for predicting customer behavior is very important for real life market and competition and it is essential to manage it. In the industry of information and communication technology [36]
NN-GA (Neural Networks-Genetic Algorithm)	NN are becoming of significant importance in automatic diagnosis and prognoses of different diseases. GA used to find the optimum network structure with high classification accuracy [37, 61–63]
HIDSS (Hybrid Intelligent Decision Support Systems)	It is implemented in an agent-based architecture, a case study system European Monetary Union-HIDSS is built for risk analysis and prediction of evolving economic clusters in Europe [46]. Neural network and rule-based agents are evolved from incoming data and expert knowledge if a decision making process requires this
gHFNN (Genetically Optimized Hybrid Fuzzy Neural Networks)	It is used for modeling vehicle for nonlinear and complex systems. It is constructed by combining Fuzzy NNs with polynomial NNs. Rule-based FNN is optimized by genetic algorithms and back-propagation (BP) learning algorithm: GAs leads to the auto-tuning of vertexes of membership function, while the BP algorithm helps produce optimal parameters of the consequent polynomial of fuzzy rules through learning. Polynomial NN that is the consequent structure of the gHFNN [38]

Benxian Yue et al., investigated optimal reducts using a particle swarm optimization approach. This approach observed change of positive region as particles proceed throughout search space is best attribute [50].

Dong-Hwa Kim et al., proposed hybrid approach genetic algorithms and bacterial foraging algorithms for function optimization problems [51]. Performance of hybrid approach is studied on mutation, chemotactic steps, crossover, variation of step sizes and lifetime of bacteria.

Various applications of the hybrid systems is given in Table 4.

3 Classification Modeling Using CI Techniques

Fuzzy and rough sets are well known for analyzing different aspects of uncertainty. Combination of these techniques will enable to build a robust mathematical approach for combating problems of uncertainty. This work suggests fuzzy-rough theory based solution for classifying medical datasets. Medical datasets are considered as special domain as most of the times, medical data is incomplete, vague and noisy. Fuzzy-rough rule induction (FRRI) is developed to generate rules for classification. Further, proposed FRRI is tested to understand the performance on

Table 5 Performance measures

Metrics	Formula	Evaluation process
Accuracy	$\sum_{i=1}^{l} \dfrac{\dfrac{tp_i+tn_i}{tp_i+fn_i+fp_i+tn_i}}{l}$	Average per-class effectiveness of a classifier
Error rate	$\sum_{i=1}^{l} \dfrac{\dfrac{fp_i+fn_i}{tp_i+fn_i+fp_i+tn_i}}{l}$	Average per-class classification error
Precision	$\dfrac{\sum_{i=1}^{l} tp_i}{\sum_{i=1}^{l} (tp_i+fp_i)}$	Agreement of the data class labels with those of a classifiers
Recall	$\dfrac{\sum_{i=1}^{l} tp_i}{\sum_{i=1}^{l} (tp_i+fn_i)}$	Effectiveness of a classifier to identify class labels

par with other CI techniques such as Multi-Layer Perceptron (MLP) and
Fuzzy-Genetic Modeling (Table 6). The metrics are used for perforamce analysis of
classifiers is shown in Table 5.

3.1 Proposed Algorithm for Rule Generation

This section gives the proposed algorithm for rule generation that uses
Fuzzy-Rough Rule Induction. In the proposed hybrid classifier lower and upper
approximation of RST is fuzzified by adopting the approach given below:

- A fuzzy set in X, is simplified to set A, allowing these objects can fit to class
  label to different degrees.
- Alternatively, estimate objects indiscernibility, their approximate equality, represented by a fuzzy relation R may be measured. Subsequently, objects are
  classified into classes with "soft" boundaries based on their equality to each
  other.

By definition, abrupt transitions between classes are adjusted by gradual ones,
allowing that an element can fit (to varying degrees) to more than one class.

Fuzzy Indiscernibility

An information system I is considered. The fuzzy indiscernibility relation R_a is
used for any fuzzy relation that determines degree to which two objects are
indiscernible. The following equations are based on a quantitative attribute in tolerance relations R_a:

$$R_a(x, y) = \max \left(\min \left(\frac{a(y) - a(x) + \sigma_a}{\sigma_a}, \frac{a(x) - a(y) + \sigma_a}{\sigma_a} \right), 0 \right) \qquad (3)$$

where $\forall x, y \in U$ with σ_a denoting the standard deviation of a. Fuzzy-rough set
theory lower and upper approximations are defined by an implicator I and t-norm τ.

The following are the fuzzy B-lower and B-upper approximations of a fuzzy set A in U.

$$(R_B\!\downarrow\!A)(y) = \mathop{\inf}_{x\,\in\,U} \tau(R_B(x,y)A(x)) \tag{4}$$

$$(R_B\!\uparrow\!A)(y) = \mathop{\sup}_{x\,\in\,U} \tau(R_B(x,y)A(x)) \tag{5}$$

$R_B \downarrow A$ is set of elements necessarily satisfying concept (strong membership), while $R_B \uparrow A$ is set of elements possibly belonging to concept (weak membership). Mainly, these were designed to deal with uncertainty in data. Fuzzy B-positive region is defined based on fuzzy B-indiscernibility relations, for $y \in X$.

$$POS_B(y) = \left(\bigcup_{x\,\in\,U} R_B \downarrow R_d x\right)(y) \tag{6}$$

Degree of dependency of d on B, γ_B by

$$\gamma_B = \frac{|POS_B|}{|U|} = \frac{\sum_{x\,\in\,X} POS_B(x)}{|U|} \tag{7}$$

An algorithm for classification of medical dataset with a hybrid approach using fuzzy-rough set theory is represented as Fuzzy-Rough Rule Induction (FRRI).

Algorithm: Fuzzy-Rough Rule Induction (FRRI)

Given medical dataset M, select randomly a subset B of conditional attribute A.

(a) Initially subsets B of conditional attribute, ruleset R and cover set Cov are empty.

(b) For each attribute a ∈ M, repeat the following steps.

(c) For each object o_1 ∈ M, repeat the following steps.

(d) Compute γ degree D_1 of belongingness of o_1 to positive region of attribute a.

(e) Compute γ degree D_2 of belongingness of o_1 to positive region for given dataset M.

(f) If degree D_1 equals to D_2.

(g) Then construct the rule r, for the object o_1 and attribute subset $B \cup a$.

(h) Add the rule to ruleset R if r does not have same or more coverage than existing rules in ruleset R. Update the coverage set.

This algorithm generates fuzzy-rough if-then rules for classifying a given medical dataset. However, it is observed that in such rules, antecedents are covering almost all attributes of decision system which eventually increase computational time in classification. It is favorable to combine rule induction and attribute selection process. Thereby, fuzzy rules are generated from best attributes that maximally cover the decision system.

3.2 Multi-layer Perceptron for Classification

Multilayer perceptron (MLP) classifier is based on the feed forward artificial neural network [32]. Here, information moves in only forward direction from input nodes and passed through next hidden nodes to output nodes. There are no cycles or loops in the network. It consists of multiple layers of nodes. Each layer is fully connected to the next layer in the network. In input layer, nodes represent input data. All other nodes map inputs to outputs by a linear combination of inputs with the node's weights w and bias b by applying an activation function. This can be written in matrix form for MLP with $K + 1$ layers as follows:

$$y(X) = f_K \left(\cdots f_2 \left(w_2^T f_1 \left(w_1^T X + b_1 \right) + b_2 \right) \cdots + b_K \right) \qquad (8)$$

A multi-layer neural network can compute a continuous output instead of real numbers called step function. A common choice is the so-called logistic function. Sigmoid function refers to the special case of the logistic function. Nodes in intermediate layers use sigmoid (logistic) function $f(Z_i)$.

These nodes of networks consists of multiple layers are interconnected in a feed-forward direction. Every neuron in one layer shown feed-forward directed network to its subsequent layer. This process is implemented by sigmoid function as an activation function in this network.

$$f(Z_i) = \frac{1}{1 + e^{-Z_i}} \qquad (9)$$

Nodes in the output layer use softmax function:

$$f(Z_i) = \frac{e^{Z_i}}{\sum_{K=1}^{N} e^{Z_K}} \qquad (10)$$

Number of nodes in the output layer corresponds to the number of classes.

3.3 Genetic-Fuzzy Modeling (GFM)

This section gives a hybrid CI approach that have attracted a lot of attention in past decade, which successfully offered solutions to many real world problems. In this work, hybridization of fuzzy logical and genetic algorithm (GA) is used to construct the classification model. GA are stochastic search based on natural selection and genetics. The fitness function simply defined as candidate solution to the problem as input and produces as output how "fit" our how "good" the solution is with respect to the problem in consideration. Calculation of fitness value is done repeatedly in a GA and therefore it should be sufficiently fast. The significant attributes selected from dataset helps in diagnosing system to built a classification fuzzy inference model. The rules for the fuzzy system are generated from dataset. These rule sets are most significant and optimal subset of rules are selected using genetic algorithm. The benefits of GA and fuzzy inference system for effective prediction of heart disease in patients. Genetic-Fuzzy Logic (GFL) model for effective heart disease prediction [33].

Genetic-Fuzzy Logic

GFL finds fitness of a chromosome in a population by genetic algorithm, as it decides termination criterion. The attributes in the dataset are selected using GA and the fuzzy inference system for classification. The fitness function value is the measurement that helps to check the nearness of the optimal solution. Genetic operators used in genetic algorithms are analogous to those in the real life: survival of the fittest, or selection; reproduction (crossover, also called recombination); and mutation.

Selection is a process of selecting parents among the population by using roulette wheel selection. Intermediate crossover is used to select parents and interchanges position of values based on crossover point fixed. The values before fixed point from one chromosome is replaced with first part of new chromosome and the values that are in the second chromosome are replaced with new ones, thus inheriting the features of both the parents. Gaussian mutation is a process of changing gene values based on its given probability 0.05 and 1. The stochastic search of the genetic algorithm stops based on the convergence criteria. GA combine high performance notions to achieve better performance for getting optimal solution.

Membership Functions

Membership function represents the fuzzy set and measures degree of similarity. It is defined as fuzzy set A on the universe of discourse X is defined as $\mu_A : X \rightarrow [0, 1]$, where each element of X is mapped to a value between 0 and 1. This value, called membership value or degree of membership, quantifies the grade of membership of the element in X to the fuzzy set A.

$$\mu_A(x, c, s, m) = exp\left[-\frac{1}{2}\left|\frac{x-c}{s}\right|^m \right] \tag{11}$$

where c is center, s is width and m is fuzzification factor.

Fuzzy Inference System

A fuzzy inference system maps inputs to output using predefined fuzzy rules available in knowledge base. The knowledge base consists of if-then fuzzy rules that specify the relationship between the input and output fuzzy sets. If

$a1, a2, \ldots, an$ are the attributes and $c1, c2, \ldots, cm$ are class labels then a fuzzy rules are based on the linguistic values. As it requires the input in fuzzy values, the input is fuzzified and output from the inference system is defuzzified.

Fuzzy Classifier

Fuzzy classifier learns data in the form of rules and predicts target value for set of test data. It uses cross validation tenfold technique for estimating the test error. This process identifies subset for testing with all the other subsets as training subsets. Using trained model testing subset is classified. It continues for ten times with different training and testing data. The actual labels are provided based on the data in the dataset whereas the modified labels are framed based on the fuzzy classifier. Here, defuzzification methods replaces fuzzy values to their corresponding crisp values by using centroid method.

In GFL, genetic algorithm implements stochastic search on dataset to reduce number of features. Fuzzy inference system predicts test data by fuzzy Gaussian membership function and centroid defuzzification method.

3.4 Rule Classifier Using Fuzzy Ant Colony Optimization

Ant Colony Optimization (ACO), an inspired algorithm from nature, has been successfully applied to classification tasks of data mining [54–56]. A rule-based system for medical data mining by using a combination of ACO and fuzzy set theory, called Fuzzy ACO-Miner was proposed by Mostafa Fathi Ganji and Mohammad Saniee Abadeh [52].

Fuzzy ACO-Miner operates in rule generations and its optimization phases. Initially, an ACO algorithm is applied to learn fuzzy rules. This algorithm applies the artificial ants to explore among the training samples and gradually deriving fuzzy rules. The ants learns the rules related to each class separately corresponding fuzzy rules.

Ant constructs rule randomly by adding one term at a time and in the next iterations the ants modify rule. Each ant chooses term to modify (or add to current rule in the first iteration) with following probability:

$$P_{i,j} = \frac{\tau_{i,j}(t) \cdot \eta_{i,j}}{\sum\limits_{i}^{a} \sum\limits_{j}^{b_i} \tau_{i,j}(t) \cdot \eta_{i,j}} \tag{12}$$

$\eta_{i,j}$ is a problem-dependent heuristic value for term. The function that defines the problem-dependent heuristic value. $\tau_{i,j}$ is the amount of pheromone currently available (at time t) between attributes. I is the set of attributes that are not yet used by the ant.

It is necessary to decrease the pheromone of terms that have not participated in the construction of rules. For this purpose, pheromone evaporation is simulated. To simulate the pheromone evaporation in real ant colony, the amount of pheromone associated with each term that does not occur in the constructed rule must be decreased. The pheromone of unused terms is decreased by dividing the amount of the value of each $\tau_{i,j}$ by the summation of all $\tau_{i,j}$ [53].

Fuzzy ACO-Miner has also have additional features that make it different from existing classifiers based on ACO meta-heuristic. Unknown data is classified using fuzzy ACO-Miner based on averaging both the number of rules and the covering value to classify the input data.

3.5 Fuzzy Discrete Particle Swarm Optimization Classifier for Rule Classification

In contrast to traditional mining approaches, biologically inspired algorithms such as evolutionary algorithms and swarm intelligence approaches can also be used for the purpose of classification task. Hence, fuzzy discrete particle swarm optimization classifier by local search (fuzzy DPSO-LS) classifier is proposed to handle imprecise and uncertain data is explained by Min Chen and Simone A. Ludwig [57].

FDPSO-LS classifier, which uses a rule base to represent a 'particle' that evolves rule base over time. It is implemented as a matrix of rules, representing fuzzy IF-THEN classification rules, that have conjunctive antecedents and one consequent. It is applied to both discrete and continuous data sets [58, 59]. In addition, a local mutation search strategy was incorporated in order to take care of the premature convergence of PSO. As the number of rules rises, an efficient algorithm that can automatically find the fuzzy rules is important and necessary. Normally, several rules of the rule base are fired in the fuzzy rule classification system. The predicted class for a given instance is determined by the membership degree of the input variables. Specifically, for each class k,

$$\beta_{\text{Classk}} = \arg \max_{k} \sum_{1 \leq i \leq n} \prod_{1 \leq j \leq m} \mu_{ij} \qquad (13)$$

where μ_{ij} is the input membership degree of the ith rule of the jth antecedent. The class that has the largest β value is selected as the predicted class. Limitation of this approach are, it may not efficiently normalize discrete datasets using linguistic terms, leads to overfitting of the model and may result in decrease of classification of accuracy.

4 Experimental Evaluation of the Classification Model

The degree of belief on classifier's capability to classify the unknown instances can be acknowledged by looking at its generalization ability on trained instances. The classifier learned on training dataset has to be tested experimentally for its performance. Experimental procedures adopted for evaluation of the proposed classification model and analysis of results is shown in this section. The experimental process is carried out by using various classification evaluation metrics for assessing the classifier's performance on a dataset.

4.1 Evaluation Metrics

Most of the study on medical data analysis is done using binary classification models. This study focuses on medical data analysis with multiclass classification problem, as most in most of the cases the automated medical diagnosis cannot be formulated as binary classification problems. In multiclass classification, the instances are to be classified into only one class from set of non-overlapping classes. Subsequently, a fuzzy-rough classifier for classification of medical datasets having uncertain data is developed. The experimental evaluation process involves in comparison of proposed classification algorithms performance with existing computational intelligence techniques.

The choice of appropriate metrics for evaluation of the classifiers is vital to measure the actual performance. Here, the focus should be on performance but not the perception of the trained classifier. The most common metric to evaluate classifiers are accuracy and error-rate. Accuracy tells the ratio of instances the classifier classified correctly. Error-rate specifies the ratio of incorrect classification done by classifier. Although, specificity and sensitivity are the popular metrics to indicate success in medical diagnosis classification models along with medical test, they are not preferred here as the experimental evaluation is done for multiclass classification problems. Precision and Recall are known to be quality metrics for multiclass classification problems. Precision specifies classifiers ability in precisely (exactness) identifying the relevant class labels by the classifier for each class. Recall specifies completeness of the classifier.

Also performance of the proposed approach is given by comparing the results obtained with that of existing well known methods.

4.2 Description of the Dataset Used

The main focus is to analyze uncertainty in medical dataset, so that efficient diagnosis of diseases can be made by medical practitioners. This study made use of

Cleveland heart disease dataset retrieved from UCI machine learning repository [34]. The dataset has 303 instances with 14 attribute. The attribute indicating level of heart disease is considered as response attribute which is distributed into five classes.

4.3 Experimental Analysis

In the experimental evaluation process, Cleveland heart disease dataset is randomly partitioned into training and testing sets. Holdout method approach is used in splitting training and test sets with 2/3, 1/3 of total instances respectively. More instances of each class can make the classifier to attain good generalization capabilities. Training dataset is given as input to the classifiers to build classification model. For evaluation of the classifier the test set is used. The results obtained are showcased subsequently. Also, a performance of the proposed classifier is evaluated by comparing results obtained with that of the other classifier' built using computational intelligence techniques on the same dataset.

Fuzzy-Rough Rule Induction

Rule based classifiers are known for their better representation and interpretation of classification model. A rule based classifier (FRRI) is built on Cleveland heart

Table 6 Classification performance

Classification models	No. of rules	Precision	Recall	Accuracy	Error rate
MLP	8	0.781	0.846	0.829	0.171
GFM	10	0.854	0.867	0.868	0.132
FACO-Miner	5	0.851	0.782	0.752	0.248
Fuzzy DPSO	7	0.799	0.751	0.748	0.252
Proposed FRRI	4	0.651	0.772	0.908	0.092

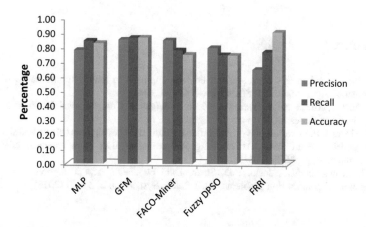

Fig. 3 Classification models

disease dataset. The proposed FRRI has generated 4 rules. Classification performed by these rules has given an accuracy of 90.8%. The performance of FRRI is tested with other CI techniques. The techniques used in comparative analysis are, Multi-layer perceptron and also, fuzzy based hybrid classification techniques: genetic-fuzzy modeling (GFM), fuzzy ant colony optimization (FACO-Miner), fuzzy particle swarm optimization (Fuzzy DPSO). It is observed that results are inferior to that of FRRI.

Table 6 depicts the classification performances of aforementioned computational intelligence techniques.

The graphical representation of classification models are depicted in Fig. 3.

5   Conclusions and Future Research Directions

The problem of analyzing medical data for better diagnosis by using computational intelligence techniques has been studied. This work emphasized on analysing uncertainty in medical data which is inevitable while collecting information from patients. A predictive hybrid model is built using fuzzy-rough classifiers that classifies a new instance/record of a patient and labels/conveys whether the patient might suffer from the disease or not (this research is carried on heart dataset). To achieve this, the model is trained to generate rules for classification. Fuzzy-rough concepts are used for the entire process as it is proven to handle uncertainty efficiently. A rule based fuzzy-rough classifier FRRI, which classifies uncertain medical datasets has been built.

Similarly, the framework can be extended to different domains other than medical field. The behavior of a netizen can be analyzed by such hybrid models. Also, the impact of social media can also be studied. Computational intelligence techniques can be used to build robust decision making systems. Further, ensemble of random forests can be used for classification. Parallel processing can be adopted to reduce the computational complexity of random forests.

References

1. G. Selvachandran, A vague soft set theoretic approach to multi attribute decision making problems. Appl. Math. Sci. **8**(134), 6937–6949 (2014)
2. J. Fulcher, *Advances in Applied Artificial Intelligence*. Computational Intelligence and its Applications Series (2006)
3. W. Sibanda, P. Pretorius, Novel application of multi-layer perceptrons (MLP) neural networks to model HIV in South Africa using seroprevalence data from antenatal clinics. Int. J. Comput. Appl. **35**(5) (2011)
4. A.J. Umbarkar, M.S. Joshi, P.D. Sheth, Dual population genetic algorithm for solving constrained optimization problems. Int. J. Intell. Syst. Appl. **2**, 34–40 (2015)

5. M. Kumar, N. Yadav, Fuzzy rough sets and its application in data mining field. Adv. Comput. Sci. Inf. Technol. **2**(3), 237–240 (2015)
6. A. Umut, S. Ayberk, Failure mode and effects analysis under uncertainty: a literature review and tutorial. Intell. Decis. Making Qual. Manage. **97**, 265–325 (2016)
7. C.M.D. Cornelis, E. Cock, E. Kerre, Intuitionistic fuzzy rough sets: at the crossroads of imperfect Knowledge. Expert Syst. **20**(5), 260–270 (2003)
8. J. David Hunter, Uncertainty in the Era of Precision Medicine. The New England Journal of Medicine. pp. 711–713, (2016)
9. J. Kacprzyk, S. Zadrozny, G. De Tre, Fuzziness in database management systems. Fuzzy Sets Syst. **281**, 300–307 (2015)
10. B. Zhang, A new measure of similarity between vague sets, in *International Conference on Oriental Thinking and Fuzzy Logic*, vol. 443 (2016), pp. 601–610
11. G. De Tre, S. Zadrożny, *Soft Computing in Database and Information Management*. Springer Handbook of Computational Intelligence (2015)
12. G.J. Klir, *Uncertainty and Information: Foundations of Generalized Information Theory* (Wiley, Hoboken, 2006), p. 499
13. L. Zadeh, *Fuzzy Sets, Information and Control* (1965), pp. 338–353
14. Z. Pawlak, Rough sets. Int. J. Comput. Inf. Sci. **11**(5) (1982)
15. I. Duntsch, G. Gediga, Rough set data analysis. Encycl. Comput. Sci. Technol. **43**(28), 281–301 (2000)
16. E.H. Shortliffe, J.J. Cimino, Biomedical informatics. Comput. Appl. Health Care and Biomed. (2014)
17. A. Skowron, S.K. Pal, Rough sets. Pattern recognition and data mining. Pattern Recogn. Lett **24**(6), 829–933 (2003)
18. S. Udhayakumar, H. Hannah, Inbarani: a novel neighborhood rough set based classification approach for medical diagnosis. Proc. Comput. Sci. **47**, 351–359 (2015)
19. Z. Pawlak, *Rough Sets: Theoretical Aspects of Reasoning About Data* (Dordrecht Kluwer Academic, 1991)
20. A. Skowron, Z. Pawlak, J. Komorowski, L. Polkowski, A rough set perspective on data and knowledge, in *Handbook of Data Mining and Knowledge Discovery* (2002), pp. 134–149
21. D. Jianhua, Rough set approach to incomplete numerical data. Inf. Sci. **241**, 43–57 (2013)
22. D.S. Yeung, D. Chen, E.C.C. Tsang, J.W.T. Lee, W. Xizhao, On the generalization of fuzzy rough sets. IEEE Trans. Fuzzy Syst. **13**(3), 43–361 (2005)
23. I. Masahiro, W.-Z. Wu, C. Cornelis, *Fuzzy-Rough Hybridization*. Springer Handbook of Computational Intelligence (2015), pp. 425–451
24. W.Z. Wu, J.S. Mi, W.X. Zhang, Generalized fuzzy rough sets. Inf. Sci. **151**, 263–282 (2003)
25. D. Boixader, J. Jacas, J. Recasens, Upper and lower approximations of fuzzy sets. Int. J. Gen. Sys. **29**(4), 555–568 (2000)
26. E. Saleh, A. Valls, A. Moreno, P. Romero-Aroca, S. dela Riva-Fernandez, R. Sagarra Alamo, Diabetic retinopathy risk estimation using fuzzy rules on electronic health record data. Model. Decis. Artif. Intell. **9880**, 263–274 (2016)
27. S. Chimphlee, N. Salim, M.S.B. Ngadiman, W. Chimphlee, S. Srinoy, Independent component analysis and rough fuzzy based approach to web usage mining, in *Proceedings of the Artificial Intelligence and Applications* (2006), pp. 422–427
28. L. D'eer, N. Verbiest, C. Cornelis, L. Godo, A comprehensive study of implicator–conjunctor-based and noise-tolerant fuzzy rough sets: definitions, properties and robustness analysis. Fuzzy Sets Syst. (2014)
29. S.P. Tiwari, S. Sharan, V.K. Yadav, Fuzzy closure spaces vs. fuzzy rough sets. Fuzzy Inf. Eng. **6**(1), 93–100 (2014)
30. J. Zhan, K. Zhu, A novel soft rough fuzzy set: z-soft rough fuzzy ideals of hemirings and corresponding decision making. Soft Comput. 1–14 (2016)

31. S. Rajasekaran, G.A. Vijayalakshmi Pai: Neural networks, fuzzy logic and genetic algorithms synthesis and applications. PHI Learn. (2011)
32. A. Rajendra, S. Priti Srinivas, Artificial neural network. Intell. Tech. Data Sci. 125–155 (2016)
33. T. Santhanam, E.P. Ephzibah, Heart disease prediction using hybrid genetic fuzzy model. Indian J. Sci. Technol. **8**(9), 797–803 (2015)
34. UC Irvine Machine Learning Repository, http://archive.ics.uci.edu/ml
35. A.P. Markopoulos, W. Habrat, N.I. Galanis, Modelling and optimization of machining with the use of statistical methods and soft computing. Des. Exp. Prod. Eng. 39–88 (2016)
36. A. Hudaib, R. Dannoun, O. Harfoushi, R. Obiedat, H. Faris, Hybrid data mining models for predicting customer churn. Int. J. Commun. Network Syst. Sci. **8**, 91–96 (2015)
37. Md. Mijanur Rahman, T. Akter Setu, An implementation for combining neural networks and genetic algorithms. Int. J. Comput. Sci. Technol. **6**(3) (2015)
38. S.-K. Oh, W. Pedrycz, Genetically optimized hybrid fuzzy neural networks: analysis and design of rule-based multi-layer perceptron architectures, in *Engineering Evolutionary Intelligent Systems* (Springer, 2008)
39. K.-L. Du, M.N.S. Swamy, *Genetic Algorithms. Search and Optimization by Metaheuristics* (2016), pp. 37–69
40. J.H. Holland, *Adaptation in Natural and Artificial Systems* (University of Michigan, 1975)
41. D.E. Goldberg, K. Sastry, A practical schema theorem for genetic algorithm design and tuning, in *Proceedings of the Genetic and Evolutionary Computation Conference* (2001), pp. 328–335
42. D.H. Kraft, E. Colvin, G. Bordogna, Fuzzy information retrieval systems: a historical perspective, in *Fifty Years of Fuzzy Logic and its Applications*, vol. 326, pp. 267–296 (2015)
43. D. Dubois, H. Prade, *Putting Rough Sets and Fuzzy Sets Together* (1992), pp. 203–232
44. T. Beaubouef, F. Petry, Information systems uncertainty design and implementation combining: rough, fuzzy and intuitionistic approaches. Flexible approaches in data. Inf. Knowl. Manage. **497**, 143–164 (2014)
45. M. Cai, Q. Li, G. Lang, Shadowed sets of dynamic fuzzy sets. Granular Comput. 1–10 (2016)
46. Kaklauskas Arturas, Intelligent decision support systems. Biometric Intell. Decis. Making Support **81**, 31–85 (2015)
47. G.M. Bergman, Ordered sets, induction, and the axiom of choice, in *An Invitation to General Algebra and Universal Constructions* (2015), pp. 119–171
48. M.H.A. Elhebir, *Machine Learning Methods for Mining Web Access Patterns* (Sudan University for Science & Technology, 2016)
49. A. Abraham, C. Grosan, V. Ramos, *Swarm Intelligence and Data Mining*. Studies in Computational Intelligence (Springer, 2006), p. 270
50. B. Yue, W. Yao, A. Abraham, H. Liu, A new rough set reduct algorithm based on particle swarm optimization, in *International Work-Conference on the Interplay Between Natural and Artificial Computation* (2007), pp. 397–406
51. D.H. Kim, A. Abraham, J.H. Cho, Hybrid genetic algorithm and bacterial foraging approach for global optimization. Inf. Sci. **177**(18), 3918–3937 (2007)
52. M.F. Ganji, M.S. Abadeh, *Parallel Fuzzy Rule Learning Using an ACO-Based Algorithm for Medical Data Mining* (IEEE, 2010)
53. R.S. Parpinelli, H.S. Lopes, A.A. Freitas, Data mining with an ant colony optimization algorithm. IEEE Trans. Evol. Comput. **6**, 321–332 (2002)
54. K.M. Salama, A.M. Abdelbar, Learning neural network structures with ant colony algorithms. Swarm Intell. **9**, 229–265 (2015)
55. P. Mrutyunjaya, A. Ajith, Hybrid evolutionary algorithms for classification data mining. Neural Comput. Appl. **26**, 507–523 (2015)

56. A. Thannob, S. Siriporn, L. Chidchanok, Optimizing the modified fuzzy ant-miner for efficient medical diagnosis. Appl. Intell. **37**, 357–376 (2012)
57. M. Chen, S.A. Ludwig, A fuzzy discrete particle swarm optimization classifier for rule classification. Int. J. Hybrid Intell. Syst. **11**(3) (2014)
58. H. Ishibuchi, T. Nakashima, T. Murata, Performance evaluation of fuzzy classifier systems for multidimensional pattern classification problems. IEEE Trans. Syst. Man Cybern. **29**(5), 601–618 (1999)
59. L.X. Wang, J.M. Mendel, Generating fuzzy rules by learning from examples. IEEE Trans. Syst. Man Cybern. **22**(6) (1992)
60. Medical Misdiagnosis in America 2008, http://www.premerus.com/news/Misdiagnosis_in_America.pdf
61. O.W. Samuel, G.M. Asogbon, A.K. Sangaiah, P. Fang, G. Li, An integrated decision support system based on ANN and Fuzzy_AHP for heart failure risk prediction. Expert Syst. Appl. **68**, 163–172 (2017)
62. A.K. Sangaiah, A.K. Thangavelu, X.Z. Gao, N. Anbazhagan, M.S. Durai, An ANFIS approach for evaluation of team-level service climate in GSD projects using Taguchi-genetic learning algorithm. Appl. Soft Comput. **30**, 628–635 (2015)
63. A.K. Sangaiah, A.K. Thangavelu. An adaptive neuro-fuzzy approach to evaluation of team-level service climate in GSD projects. Neural Comput. Appl. **23**(8) (2013). doi:10.1007/s00521-013-1521-9

Index

© Springer International Publishing AG 2017
A.K. Sangaiah et al. (eds.), *Intelligent Decision Support Systems for Sustainable Computing*, Studies in Computational Intelligence 705, DOI 10.1007/978-3-319-53153-3

Printed in the United States
By Bookmasters